CHEMISTRY RESEARCH AND APPLICATIONS

THERMAL DECOMPOSITION

PROCESS AND EFFECTS

CHEMISTRY RESEARCH AND APPLICATIONS

Additional books and e-books in this series can be found on Nova's website under the Series tab.

CHEMISTRY RESEARCH AND APPLICATIONS

THERMAL DECOMPOSITION

PROCESS AND EFFECTS

KALYNA JOCOSA PACE
EDITOR

nova
science publishers
New York

NOTICE TO THE READER

The Publisher has taken reasonable care in the preparation of this book, but makes no expressed or implied warranty of any kind and assumes no responsibility for any errors or omissions. No liability is assumed for incidental or consequential damages in connection with or arising out of information contained in this book. The Publisher shall not be liable for any special, consequential, or exemplary damages resulting, in whole or in part, from the readers' use of, or reliance upon, this material. Any parts of this book based on government reports are so indicated and copyright is claimed for those parts to the extent applicable to compilations of such works.

Independent verification should be sought for any data, advice or recommendations contained in this book. In addition, no responsibility is assumed by the Publisher for any injury and/or damage to persons or property arising from any methods, products, instructions, ideas or otherwise contained in this publication.

This publication is designed to provide accurate and authoritative information with regard to the subject matter covered herein. It is sold with the clear understanding that the Publisher is not engaged in rendering legal or any other professional services. If legal or any other expert assistance is required, the services of a competent person should be sought. FROM A DECLARATION OF PARTICIPANTS JOINTLY ADOPTED BY A COMMITTEE OF THE AMERICAN BAR ASSOCIATION AND A COMMITTEE OF PUBLISHERS.

Additional color graphics may be available in the e-book version of this book.

Library of Congress Cataloging-in-Publication Data

ISBN: 978-1-53615-460-3

Published by Nova Science Publishers, Inc. † New York

CONTENTS

PREFACE

Azide compounds are high energy materials with applications in many chemical reactions and applications. In this compilation, several methods are applied to calculate the thermal decomposition properties of azide compounds comprising of differential scanning calorimetery, differential thermal analysis and thermogravimeteric analysis which demonstrate temperature, energy and weight loss.

Following this, the authors assess the catalytic efficacy of nanosized bimetallic transition metal oxides of copper, iron and cobalt, with iron synthesized using the solution combustion method in a laboratory reactor. The prepared catalysts are characterized for their structure and particle size.

The authors also discuss thermal runaway in alkaline batteries. It is shown that a lot of experimental data exists that contradicts the generally accepted thermal runaway mechanism.

In conclusion, a novel kinetic model for pyrolysis of medium-density fiberboard derived Beech sawdust is proposed. The complete procedure includes a combined four parallel reaction model and three-component Log-Normal distributed reactivity model.

Chapter 1 - Azide compounds are high energy materials that have been applied in many chemical reactions and applications. There are two groups of azides including organic and inorganic, and their thermal decomposiiton has been discussed. Several methods have been applied to calculate the thermal decomposition properties of azide compounds comprising differential scanning calorimetery (DSC), differential thermal analysis (DTA) and thermogravimeteric analysis (TGA) which show the temperature, energy and weight loss during the decomposition. In this chapter, a detailed discussion on thermal decomposition of various azide compounds including sodium azide, barium azide, lead azide, silver azide and some organic azides have been discussed.

Chapter 2 - Thermal decomposition studies have been carried out on nano-catalyzed compositions of AP, HTPB and its propellants and compared with that of pure AP, HTPB and uncatalyzed HTPB-AP propellants. Iron oxide and copper chromite/oxide have been extensively used as burn rate modifiers for composite solid propellants. This study aims to assess the catalytic efficacy of nanosized bimetallic transition metal oxides of copper - iron and cobalt - iron synthesized using the solution combustion method in a laboratory reactor. The prepared catalysts were characterized for their structure and particle size. It has been observed that addition of 3 wt. % $CoFe_2O_4$ and $CuFe_2O_4$ to AP, significantly lowers its high temperature decomposition temperature by 44.6°C and 79.3°C respectively. However, these catalysts have shown negligible influence on thermal decomposition behavior of fuel binder alone. Burn rate studies carried out in a Crawford high pressure strand burner to assess the variation of burn rate with pressure for virgin and catalyzed compositions showed that the addition of 3 wt.% of $CoFe_2O_4$ gave the highest burn rate of 16.97 mm/s at 6.205 MPa. Burn rate studies of propellants containing 3 wt. % of Fe_2O_3 and CuO were also carried out to compare the burn rate characteristics of synthesized catalysts. It can be concluded that $CoFe_2O_4$ appears to be a better burn rate enhancer than commercial Fe_2O_3 which is generally used in composite solid propellant formulations. The combustion index, however, is found to increase with the increase in the concentration of $CoFe_2O_4$ in propellant composition. The composition giving the highest burn rate also exhibits a higher value of combustion index of the order of 0.66. Addition of aluminum powder along with meagre amounts of activated charcoal and $CoFe_2O_4$ in a formulation is seen to give a 30%

increase in burn rate of the base composition at 80 wt. % solid filler loading. This study establishes that the synthesized bimetallic nano-sized TMOs can be superior burn rate modifiers in comparison to conventional TMOs.

Chapter 3 - This chapter discusses the thermal runaway in alkaline batteries. It was shown that a lot of experimental data exists that contradict the generally accepted thermal runaway mechanism. On the basis of all the available experimental data, there was proposed the new thermal runaway mechanism. There it was shown that there are two processes of accumulation that step by step bring the nickel-cadmium batteries to the thermal runaway. First, this is a process of hydrogen accumulation inside of battery electrodes during their operation. Second, this is a process of dendrites accumulation on the cadmium electrodes of the batteries. It was experimentally proved (using thermal decomposition of sintered electrodes) that during Ni-Cd batteries long service life (more than five years), the hydrogen accumulation in large quantities takes place in a form of nickel hydrides in a sintered nickel matrix of oxide-nickel electrodes. In this chapter, it was deduced from experiments that an exothermic reaction in the thermal runaway process in alkaline batteries is the electrochemical reaction of atomic hydrogen recombination $H_{ads\ Cd} + H_{ads\ Ni} \rightarrow H_2 \uparrow$ ($H_2O + H_{ads} + e^- \rightarrow H_2 \uparrow + OH^-$ on a cathode and $H_{ads} + OH^- \rightarrow H_2O + e^-$ on an anode). A rate-limiting step for this reaction on both a cathode and an anode is found to be a step of metal-hydrides disintegration. This electrochemical reaction proceeds at voltage 0.5–0.6 V on battery terminals. An analytical model of thermal runaway in alkaline batteries is proposed.

Chapter 4 - A novel kinetic model for pyrolysis of *MDF (medium-density fiberboard)* derived Beech sawdust was proposed. The complete procedure includes a combined four parallel reaction model (CPRM) and three-component Log-Normal distributed reactivity model. Based on the matrix analysis, it was found that for quality pyrolysis performing, the lower values of the heating rates (where the possibility of the secondary reactions is significantly reduced) are highly desirable. CPRM approach showed that with an increase of heating rate, pyrolysis rate of UF-resin adhesive rapidly increases and its rate curve moves very close to decomposition curve related to cellulose. At the highest heating rate, the UF-resin adhesive pyrolysis curve “overrated” cellulose-related curve. It was assumed that UF-resin adhesive participates in cellulose decomposition, hastening chemical bonds breaking in cellulose molecule, and further stimulating lignin decomposition. Three-component Log-Normal distributed reactivity model showed that certain asymmetries in effective activation energy distribution curves for some pseudo-components is a consequence of a wider variety of bonds present, and/or “capacity” in the tendency of an easier bond breaking. Both approaches, contained in the newly proposed model, confirm the strong deviation from the additive behavior (the adhesive compound present in the studied lignocellulose material does not produce the synergistic effect).

In: Thermal Decomposition
Editor: Kalyna Jocosa Pace
ISBN: 978-1-53615-460-3

Chapter 1

THERMAL DECOMPOSITION OF AZIDE COMPOUNDS

Nafise Modanlou Juibari, PhD and Sara Tarighi*, PhD
Faculty of Petrochemicals,
Iran Polymer and Petrochemical Institute, Tehran, Iran

ABSTRACT

Azide compounds are high energy materials that have been applied in many chemical reactions and applications. There are two groups of azides including organic and inorganic, and their thermal decomposiiton has been discussed. Several methods have been applied to calculate the thermal decomposition properties of azide compounds comprising differential scanning calorimetery (DSC), differential thermal analysis (DTA) and thermogravimeteric analysis (TGA) which show the temperature, energy and weight loss during the decomposition. In this chapter, a detailed discussion on thermal decomposition of various azide compounds including sodium azide, barium azide, lead azide, silver azide and some organic azides have been discussed.

---

* Corresponding Author's E-mail: s.tarighi@ippi.ac.ir.

Keywords: Thermal decomposition, Organic azide, Inorganic azide, Differential scanning calorimetry, Differential thermal analysis

INTRODUCTION

Thermal Decomposition

Thermal decomposition of solids is a key phenomenon in many natural processes and industrial technologies. Karst phenomenon and volcanic activity, rock weathering, dehydration of rock-salt deposits, which are actual showing of the evolution of the Earth and of the Universe as a whole, have been subjects of interest from the ancient time. The history of our civilization is intimately relevant to the development of a variety of technologies based on heat treatment of natural materials with the purpose of obtaining new objects of practical importance.

When a solid is heated, increased vibrations of the lattice caused one (or more) of the following changes:

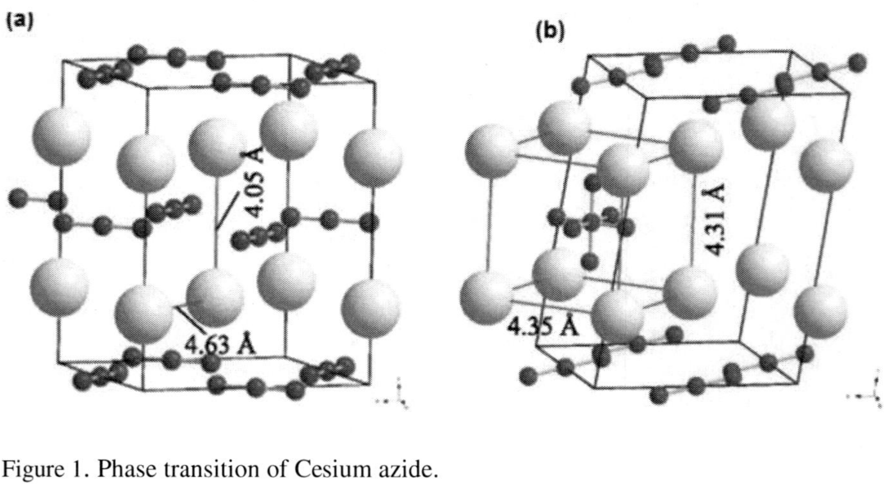

Figure 1. Phase transition of Cesium azide.

Phase Transition [1, 2]

A new arrangement of constituent molecules at higher temperatures which is stable and thereby accommodating a different pattern of the constituent movements. For example, many metal perchlorates undergo a phase transition from a rhombohedral to cubic structure [3]. This structural modification allows an increased movement of the tetrahedral ClO_4^- ion at higher temperatures. The phase transition of Cesium azide is given in Figure1.

Melting [4]

When sufficient energy is available for the molecules, attraction forces between the constituents become insufficient to maintain the ordered arrangement of the solid and the system transforms to the more disordered arrangement of constituents in a liquid. At temperatures just above the melting point, the density of the liquid is usually comparable with that of the solid from which it was formed, indicating a relatively small change in the packing of the constituents. However, the density of the most atoms in liquid form decrease rapidly with increasing the temperature that indicates the disorder development. In the case of some complex molecules, the change from solid to liquid may occur in multiple stages. The structure of intermediate levels are known as liquid crystals [4].

Sublimation

When the bonding forces in the solid are rapidly overwhelmed by the kinetic energy of the constituents, direct transition to the disordered arrangement of a gas may occur, without the intervention of a liquid phase.

Thermal Decomposition

When the bonding forces within the constituent molecules or ions in the solid are weaker than those between the constituting atoms, increasing the temperature may result in bond redistribution and the products chemically different from the reactant will be formed. Such chemical processes are referred to thermal decomposition (or crystolysis). These events are summarized in Table 1.

Table 1. Properties of a solid after thermal events

Phase transition	A(s)(phase 1)→A(s) (phase 2)
Melting	A(s)→A(l)
Sublimation	A(s)→A(g)
Thermal decomposition	A(s)→B (s) + gas or gases

Methods of Measuring the Thermal Properties

Differential Scanning Calorimetery (DSC)

Calorimetry is the most common technique for evaluating the thermal properties of materials. It defines a connection between temperature and specific physical properties of substances and is the only method for direct determination of the enthalpy associated with the process [5, 6]. Calorimeters are frequently used in chemistry [7], biochemistry [8, 9], biotechnology[10], pharmacology [11], and in nanoscience [12] to measure the thermodynamic properties of biomolecules and the nano-sized materials. DSC acts as an apparatus to define how physical properties of a sample change with the temperature. In other words, it is a thermal analysis instrument that determines the temperature and heat flow associated with the material transitions as a function of time and temperature [13]. During changes in temperature, DSC measures the amount of heat radiated or absorbed by the sample based on the temperature difference between the sample and the reference material [13].

Based on the operation mechanism, DSCs can be classified into two categories: heat-flux DSCs and power-compensated DSCs. In a heat flux DSC, the sample material is enclosed in a pan and an empty reference pan are placed on a thermoelectric disk surrounded by a furnace. The furnace is heated at a linear heating rate and the heat is transferred to the sample and reference pan through the thermoelectric disk. However, owing to the heat capacity (Cp) of the sample, there would be a temperature difference between the sample and the reference pan, which is measured by thermocouples. The consequent heat flow is determined by the thermal equivalent of Ohm's law:

$$q = \frac{\Delta T}{R} \quad (1)$$

where q is "sample heat flow", ΔT refers to "temperature difference between sample and reference", and R is "resistance of the thermoelectric disk"[14]. In a power-compensated DSC, the sample and reference pans are placed in separate furnaces heated by separate heaters. The sample and reference are maintained at the same temperature, and the difference in thermal power required to maintain them at the same temperature is measured and plotted as a function of temperature or time.

Differential Thermal Analysis (DTA)

DTA involves heating or cooling a test sample together with an inert reference under identical conditions, while recording any temperature differences between the sample and reference. This differential temperature is then plotted against time or temperature. Changes in the sample that lead to the absorption or evolution of heat, can be detected relative to the inert reference.

Differential temperatures can also arise from two inert samples when their responses to the applied heat is not equal. Therefore, DTA can be used to study the thermal properties and phase changes which do not lead to a change in the enthalpy. The baseline of DTA curve exhibits discontinuities at the transition temperatures and the slope of the curve at any point will depend on the microstructural constitution at that temperature.

A simple DTA curve may consist of linear portions displaced from the baseline because the heat capacities and thermal conductivities of the test and reference samples are not identical. The observed peaks are corresponded to the evolution or absorption of heat related to the physical or chemical changes in the test sample.

There are difficulties with the measurement of transition temperatures using DTA curves. In principle, the onset of a DTA peak gives the start temperature, but there may be temperature retardations depending on the location of the thermocouple with respect to the reference and test sample. It is wise to calibrate the apparatus with materials of precisely known melting points. The peak area (A), which is related to enthalpy changes in the test sample is enclosed between the peak and the interpolated baseline. The differential thermocouples are not in physical contact with the test and reference materials.

Thermogravimetric Analysis (TGA)

In TGA experiments, the sample mass is recorded as a function of temperature and time while the sample is subjected to a linear heating rate or an isothermal treatment [15]. The samples (up to several mg) are filled into an open crucible made of platinum or alumina which is attached to the arm of a recording microbalance, the so-called thermobalance. The sample is heated in a temperature-controlled furnace according to a pre-programmed temperature/time profile. During the experiment, both the furnace and thermobalance are purged independently with inert gas (usually argon or nitrogen). Since the mass changes of a sample are recorded in TGA experiments, the method is predominantly used to investigate the thermal decomposition behavior of compounds. It typically provides information on the decomposition onset temperature and the mass loss that occurs during the decomposition reaction. triazidopentaerythrite acetate, for example, degrades completely during its strongly exothermic decomposition forming only gaseous products within one total mass loss step. However, other energetic azides degrade stepwise [16].

AZIDE COMPOUNDS

Organic Azides

Organic azides are considered as explosives when the azido content is remarkably high. Of course, there is no sharp threshold at which the explosive hazard starts. However, as a rule of thumb violent, the decomposition reactions are expected for azido compounds having a (C+O)/N ratio of < 3 [17]. Organic azides with high azido content are very sensitive to friction and impact, causing strong explosions. For example, cyanuric azide is very sensitive against mechanical stimulus and thus decomposes very easily by detonation. Although the initiation power of this detonation exceeds that of classical primary explosives, no technical application has been found so far for this compound due to its high vapour pressure.

More than a century after the discovery, organic azides have received renewed interest in synthetic chemistry and established as an important and versatile class of chemical compounds [18-20]. In particular, in the past two decades an increasing interest in organic azides could be observed due to their vast synthetic utility in conjunction with their easy accessibility via various synthetic routes. Organic azides are currently considered as powerful precursors for reactive species such as nitrenes and nitrenium ions as well as nitrogen-rich materials such as aziridines, azirines, triazoles, triazolines and triazenes. Moreover, they can be easily transformed into amines, isocyanates and other functional molecules and have more recently received an increasing interest as valuable and versatile reagents within the concept of 'Click Chemistry' [21, 22].

However, in conjunction with their huge utility in organic synthesis, the potential hazardous properties of organic azides must be also carefully taken into account. Organic and inorganic azides are first and foremost energy-rich molecules which often exhibit explosive properties. The azido group is a high-energy functional group. The N_3 π-bond can be easily polarized that consequently results in strong exothermic dissociation reactions. This leads to release of molecular nitrogen and reactive nitrene groups. In general, the introduction of an azido group into an organic compound increases its energy content to approximately 290 – 355 kJ mol^{-1} [23]. This is one of the reasons that organic azides are considered and partly used as energetic materials such as energetic polymers or high-energy-density-materials (HEDM) in explosives or propellant formulations [24, 25]. However, many organic compounds containing azido groups have not found wide application as practical energetic materials because of their high sensitivity to heat and shock stimuli. The poor thermal and mechanical stability of organic azides is probably the most crucial characteristic that has prevented the chemists from extending their research on azido compounds.

Organic azides also show remarkable lower ignition temperatures in comparison with the inorganic metal azides. Most of the organic azides decompose approximately at 180 °C. Some organic azides also show light sensitivity [26], and strong incompatibility with certain chemicals. Several examples have been published in which the azides were exploded when they were brought in contact with sulfuric acid or other compounds [27, 28].

Chemistry of Organic Azides

The chemistry of organic azides were commenced with the synthesis of phenyl azide by Griess in 1864 [29] and discovery of the acyl compounds rearrangement with hydrogen azide (HN_3) by Curtius in 1890 [30]. Since 1950, various synthetic organic reactions have been developed using acyl, aryl and alkyl azides which have been extensively applied for the synthesis of nitrogen-containing azaheterocycles as well as peptides [31].

Organic azides possess diverse chemical reactivities [32] Owing to their 1,3-dipole character. They undergo [3+2] cycloaddition with unsaturated bonds, such as those in alkynes and alkenes as well as carbonitriles [33]. Organic azides can also be regarded as nitrene equivalents [34]. Accordingly, their reactions with nucleophilic anions, electrophilic cations and radicals can formally provide the corresponding nitrogen anions, cations, and radicals, respectively. This results in forming a new bond with the internal azido nitrogen along with releasing molecular nitrogen. Moreover, the generation of anions, cations, and radicals at the α-position to the azido moiety can result in rapid denitrogenation to deliver the corresponding iminyl species. The iminyl molecules can be used in further synthetic transformations through carbon–nitrogen bond formation.

Vinyl Azides

The parent compound of vinyl azide has been known for about hundred years when it was prepared and characterized by Forster and Newman. However, after publication of the first example of vinyl azides, these compounds were ignored for approximately fifty years. Perhaps, the multi-step synthesis and the assumption that dehydrohalogenation of azidohaloalkanes may result in vinyl and allyl azide mixtures altered the investigation of vinyl azides.

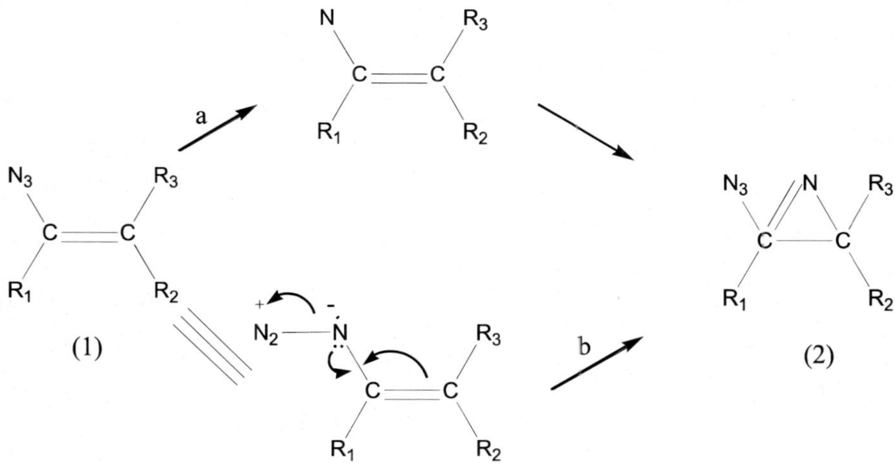

Figure 2. Mechanism of the thermal decomposition of vinyl azide.

One of the attractive chemical properties of vinyl azides is their ability to undergo thermal decomposition to give highly strained three-membered cyclic imines and 2H-azirines which are produced via vinylnitrene intermediates following the denitrogenation reaction. Moreover, the carbon–carbon double bond of vinyl azides can be used for the formation of new carbon–carbon bonds with appropriate organometallic compounds (R′–[M]) or radical species (R¢) which results in the generation of iminyl metals or iminyl radicals, respectively [35]. These iminyl species serve for the formation of carbon–nitrogen bonds. Although thermal decomposition of vinyl azides (1) gives rise to a variety of products depending on substituents, most of them are rationalized by assuming of an azirine as intermediate [36]. Two mechanistic pathways are currently given for 1 to 2 transformation. The vinyl azide can either lose nitrogen to give a vinyl nitrene which then cyclizes to the azirine (path a), or decompose with simultaneous ring closure to give 2 directly (path b) (Figure 2).

Diphenylphosphoryl Azide (DPPA)

Diphenylphosphoryl azide has been obtained by reaction of the phosphorochloridate with sodium azide. This compound undergoes pseudohalogen replacement of the azido group by treatment with nucleophilic reagents, such as water, butanol, ammonia, and various amines.

On account of the higher price of diphenylphosphoryl azide compared with trimethyl silyl azide (TMSA) and even more unfavorable ratio of azide mass in DPPA, the technical-scale use of DPPA is restricted to a few special applications such as direct conversion of carboxylic acids to the corresponding carboxylic acid azides without having to go through the activating intermediates like carboxylic acid chlorides, anhydrides or esters [37, 38]. Appling DPPA in various Mitsunobu reactions have been reported. The demand for the technical-scale quantities of DPPA will probably limited to about 10 tons per year due to the high price and the above-mentioned unfavorable mass ratio. The database of the directory of world chemical producers (DWCP) lists 15 companies in Asia, Europe and USA as producers of DPPA.

Compared to TMSA, DPPA is more safe and easy handling due to the low azide content despite the relatively high molecular weight. However, the toxicity should always be considered. DPPA can be produced via various methods which are illustrated in literatures [39]. The structure of DPPA is presented in Figure 3. When DPPA was heated to 165 ºC, the bubbling began and became vigorous, with darkening at 175 ºC. At 230 ºC, a rapid decomposition was observed and the toxic fumes of phosphorus oxides and/or phosphine may be formed. When the liquid becomes dark red-brown; the increasing heating was accompanied by progressive decrease in the size of the azide band in the Infrared spectra of the decomposition products.

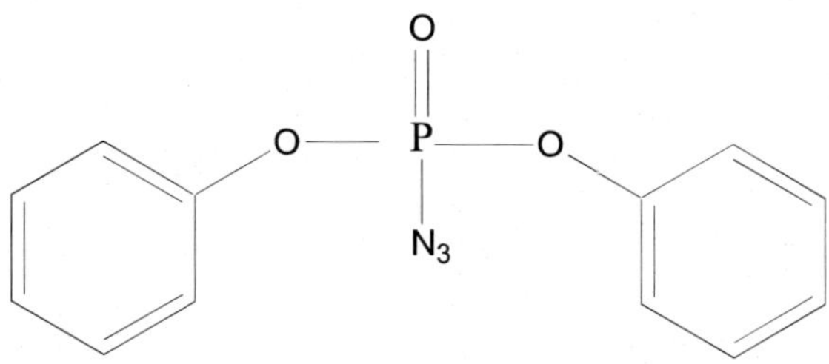

Figure 3. The chemical structure of DPPA.

Sulfonyl Azide

Sulfonyl azides, $RS(O)_2N_3$, are useful starting materials for the synthesis of biologically active enaminones, amidines, and amidates [40, 41]. Similar to the decomposition of carbonyl azides, sulfonyl azides may undergo pseudo-Curtius rearrangement into N-sulfonyl imine (RN_SO_2) via the sulfonyl nitrene intermediate, $RS(O)_2N$. However, the early solution chemistry and more recent time-resolved spectroscopy studies [42] failed to obtain convincing evidence for the formation of the final rearrangement product, $RNSO_2$, upon both the thermal-decomposition and photo-decomposition of sulfonyl azides. A number of sulfonyl nitrene intermediates had been either directly detected or chemically trapped. The decomposition of arylsulfonyl azides, $ArS(O)_2N_3$, in solution was found to be rather complex. According to the previous solution studies [43, 44] thermolysis and photolysis of arylsulfonyl azides in solvents produce traces of SO_2 and nitrene derivatives, $ArS(O)_2N$. The chemical structure of arylsulfonyl azide is shown in Figure 4.

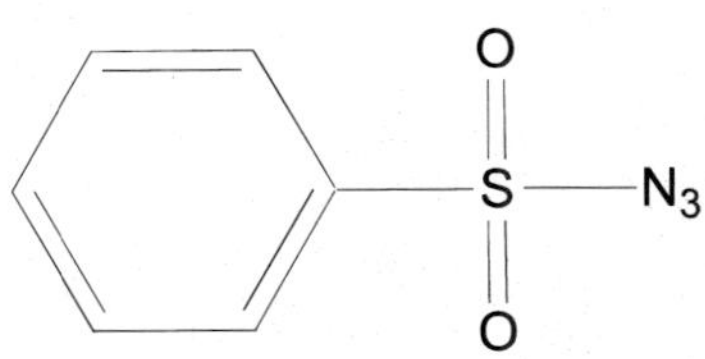

Figure 4. The chemical structure of arylsulfonyl azide.

Inorganic Azides

The inorganic azides constitute a particularly well-defined group of relatively simple ionic reactants. Their decompositions have been extensively investigated for both theoretical and practical importance. The large, relatively perfect and anhydrous crystals of many of these salts can be prepared [45], The stoichiometry of decomposition is normally simple because the anion breakdown results in the formation of gaseous nitrogen and residual metal (sometimes together with some metal nitride). Prior to 1970s, the articles related to the azide decompositions mostly contributed the theory of nucleation and growth processes with particular reference to the role of identified crystal imperfections in the generation of new reaction sites as well as interfaces. Inorganic azide decompositions is also applied as gas generators in inflatable safety cushions ("air bags") in automobiles [46]. Difficulties encountered in studying the azides as reactants include their sensitivity to light as well as hydrolysis by traces of water, and their unpredictable instability.

Metal azides run very quickly into detonation. This specific property has established the use of silver azide and lead azide as primary explosives in detonators. Remarkably more stable in terms of safe handling are lithium and sodium azide which can be more reliably used in the laboratory since they are hard to initiate explosion by impact or friction. For most laboratory conditions, alkali metal azides are not considered as explosives [47]. However, if ignited or when exposed to strong heat, alkali metal azides decompose rapidly with the evolution of large volumes of nitrogen gas.

In general, metal azides show an increase in mechanical sensitivity when going from earth alkali metal azides to heavy metal azides which are highly sensitive and explosive. The impact sensitivity of metal azides decreases in the following order: [45]

Copper > Lead, Mercury > Nickel > Cobalt > Manganese > Barium > Strontium > Calcium > Silver > Thallium > Zinc > Lithium = no explosion

However, this sequence of impact sensitivities represents only one type of mechanical stability and the order will be changed in the case of friction sensitivity. For example, silver azide is approximately 10 times more friction-sensitive than lead azide [48].

Decomposition of Inorganic Azides

The optical properties of metal azides permit the determination of the corresponded band structures. and the energy requirements for the generation of various defects can be estimated. Results obtained from the thermal studies can often be compared with observations obtained in parallel investigations of the radiolysis of azides. Additional information is also often available from measurements of the conductivities and photoconductivities of the solids, including the influences of various impurities in the crystals. Pre-irradiation generally increases the rate of subsequent thermal decomposition of azides and such studies can provide useful information on decomposition.

Imperfections are recurrent features of many proposed decomposition mechanisms which are thus usually developed in greater detail than for most other solids. The physical and chemical properties of the inorganic azides have been extensively reviewed [49, 50]. Richter [51] has discussed the chemical classification of azides as stable ionic azides, heavy-metal azides and unstable covalent azides. This classification is based on the ionic character percentage of the metal-azide bond, tabulated as formal ionicities that was previously reported [51]. For example, $Na\text{-}N_3$ and $Ba\text{-}N_3$ bonds have 70% ionic character, but $Pb\text{-}N_3$ and $H\text{-}N_3$ posses only 34% and 22% ionc character, respectively. Bertrand *et al.* [48] have reviewed the photochemical and thermal behavior of organometallic azides. Richter has also given an excellent review corresponding to the preparation methods of HN_3 and other azides. He criticizes early workers for inadequate purification and characterization of their starting materials and their negligence for the possible formation of hydrates (e.g., barium azide may be present as $Ba(N_3)_2.1.5H_2O$ below 284 K, monohydrate between 284 and 325.5 K and anhydrous form above 325.5 K). The crystal structures of the azides have been reviewed by Choi and the table of lattice dimensions was provided. He points out that the metal azide hydrates have quite different crystal structures from those of the anhydrous forms. He also notes a correlation between structure and impact sensitivity to explosion which increases with the asymmetry of the azide ion, in the following sequence:

$KN_3 < Ba(N_3)_2 < Pb(N_3)_2 < Cu(N_3)_2$

The geometric models of azide reactions are based upon the process of nucleation and growth of the product nuclei by interface. Nucleation can be defined as "the initial establishment of a new and discrete product particle within the solid reactant". Growth is thus the maintained reaction within the interface between the reactant and the product phases. The zone in which the chemical changes occur, may be regarded as an advancing energetic wave, while components of the the reactant and of the residue undergo a relatively minor displacements. Identification of the chemical changes that contribute to the nucleation process is a matter of considerable experimental difficulty. The sites of locally enhanced reactivity in which nucleation occurs are changed beyond recognition by the formation of the product particle. Characterization of these changes presents a considerable challenge to the laboratory techniques, particularly for a site embedded larger crystal [52]. The relative ease of nucleation exerts an important influence on the kinetics of the overall reaction. When the nucleation step occurs infrequently, a sigmoid yield-time curve is observed which is often ascribed to a large activation energy. In contrast, if nucleation is rapid, all surfaces of the reactant are rapidly converted to product at the early stages of the reaction. Therefore, the interface could also be formed inwards. Yield-time curves are then predominantly or entirely deceleratory.

Barium Azide

The sigmoid α-time curves observed for the decomposition of barium azide are often characteristic of nucleation and growth processes and such a mechanism was here confirmed by microscopic observations [53]. This early study by Wischin has been described as "one of the most important experimental papers on solid decompositions'. The kinetic observations include the following results: (i) the number of nuclei has been increased during the isothermal reaction with the third power of time (E~ - 310 kJ mol^{-1}); (ii) the increase in radius of each growing nuclei was directly proportional to time (Ea = 98 kJ mol^{-1}); and (iii) the rate of nitrogen evolution obeyed the power law with n = 6 to 8 and Ea = 695 kJ mol^{-1}. Mott [54] suggested that the nucleation step requires either the bimolecular combination of two interstitial barium ions, or the bimolecular trapping of two conduction electrons. The rate-determining step proposed for growth was the transfer of an electron from the azide valence band to the metal, subsequently attracting an interstitial barium ion into the developing nucleus. The positive holes (N_3) generated were diffused to the surface and reacted to form N_2. Objections to this mechanism is resulted from the inconsistencies between the measured conductivity of the solid and the calculated concentrations.

Both photolytic and thermal decompositions of barium azide were studied by Thomas and Tompkins [55]. In initial period of slow growth of nuclei, the reaction obeyed the power law $\alpha= k\ (t- t_0)^6$ and the value of Ea (602 kJ $mo1^{-1}$) was just below the value reported by Wischin (Both values were calculated incorrectly and are unreasonably large!). The proposed mechanism was the combination of mobile F-centers to form a double F-center as nucleus. Anion decomposition was proceeded at the metal interface in which the gaseous product N_2 has been evolved. The initial slow growth of nuclei was (probably incorrectly) attributed to the high value of E. Torkar *et al.* [56] confirmed the main features of the above behavior but also noted a few points of difference. They found a fourth power dependence (n = 4) of α upon time rather than n = 6 which was previously reported. Nucleation according to an exponential law occurred on (100) and (001) planes, generating randomly distributed growth particles of the product. The rates of reactions were directly proportional to the surface area of the salt. The yields of up to 73% for Ba_3N_2 were found, whereas only barium metal was reported as the product in similar reports. Salot and Waft [57] have described the preparation of barium pemitride from thermal decomposition of barium azide. The steps which culminate in the generation of a metallic nucleus during the decomposition of BaN_6 have not been completely established. Reaction probably occurs at or near the surface imperfection locations. Once established, the nucleus grows through an interfacial reaction in which the activation energy is required to transfer an electron from the valence band of the azide ion to the Fermi level of the metal. Subsequent reaction of the azide radical with another thermally excited anion yields nitrogen gas. Interaction between barium ions and the charged surface of the metallic nucleus results in incorporation of Ba atoms and growth of the nucleus, by a mechanism showing a formal similarity to electrode processes.

$$Ba(N_3)_2(s) \rightarrow Ba(s) + 3\ N_2\,(g) \quad (2)$$

Strontium Azide

α-time curves for the decomposition of strontium azide [58] are also sigmoidal with the initial stage fitted by the power law with $n = 3$. The contracting volume equation described the decay period and $E_a = 97$ kJ mol^{-1}. Reaction was suggested to involve the interaction of positive holes and excited azide ions. The decreased stability resulting from γ-preirradiation was attributed to the introduction of a high concentration of vacancies. When decomposition in vacuum was interrupted and the sample was cooled to room temperature, decomposition continued immediately, as normal, on rapid reheating of the sample to reaction temperature. If the partially decomposed reactant was exposed to water vapour after interruption, further decomposition was resembled a fresh sample. This indicates the significant role of metal nuclei which readily react with water and influence the decompositions of both irradiated and γ-irradiated salts. Studying barium and strontium azides decomposition (463 to 508 K) while being "co-irradiated" with high intensity UV [59] revealed that both thermal and photolytic decomposition occurred concurrently during their decomposition process.

Calcium Azide

The kinetics of decomposition of calcium azide is irreproducible. However, several contributory cases can be eliminated with careful control of conditions and the changes which result from aging can be investigated. Following an induction period, the acceleratory reaction can be described by the power law with n = 3. The reaction below 370 K is believed to occur through the growth of surface nuclei (E_a = 75 kJ mol^{-1}), but above this temperature, annealing could influence the magnitudes of Arrhenius parameters. The decomposition is accelerated by preirradiation [60]. The thermal decomposition of calcium azide is as following equation (eq.3).

$$Ca(N_3)_2(s) \rightarrow Ca(s) + 3\ N_2(g) \tag{3}$$

Lithium Azide

Although the rates of decomposition of lithium azides markedly differ by the preparation method, reproducible behavior was observed for the crushed and pelleted salts. Such pretreatment was believed to produce a uniform concentration of defects within the reactant assemblages. The sigmoid α-time curves fitted the Avrami-Erofeev equation with n = 3 between $0.02 < \alpha < 0.58$ and at the contracting volume expression across the wider interval $0.05 < \alpha < 0.95$, the value of E_a was 119 kJ mol^{-1}. Reaction was involved the three-dimensional growth of a constant number of nuclei and it was suggested that acceleration of rate following preirradiation resulted from an increase in the number of such nuclei [61].

Sodium Azide

Sodium azide is the most frequently used azide precursor for both laboratory scale and technical production. According to commercial production, Wislicenus developed a two-step process in 1892:

$$2Na + 2NH_3 \rightarrow 2NaNH_2 + H_2 \tag{4}$$

$$2NaNH_2 + N_2O \rightarrow NaN_3 + NaOH + NH_3 \tag{5}$$

At the first step, molten sodium is converted to sodium amide by ammonia at 350 ° C in a closed steel reactor. In second step, this product reacts with nitrous oxide to give sodium azide. The gas composition and the extreme reactivity of sodium and sodium amide with trace of water represents a particular safety risk of this method.

Variations of this production process are described. Curtius and Thiele developed another production process where a nitrite ester is converted to sodium azide using hydrazine [62]:

$$2NaNO_2 + 2C_2H_5OH + H_2SO_4 \rightarrow 2C_2H_5ONO + Na_2SO_4 + 2H_2O \quad (6)$$

$$C_2H_5ONO + N_2H_4.H_2O + NaOH \rightarrow NaN_3 + C_2H_5OH + 3H_2O \quad (7)$$

Currently, around 20 suppliers in all over the world including China, India, Japan and USA produce 1000 tons per year NaN_3. However, the data corresponding to the production procedure is rare. The major NaN_3 consumption is in the automotive industry [63]. According to the toxicity of NaN_3 and the disposal problems encountered by recycling the cars at the end of their useful life, excess capacities in future will be expected The growing need for NaN_3 in organic syntheses cannot compensate the decline in sales in the automotive sector unless substitutes such as 5-aminotetrazole has been developed.

Walker *et al.* [64] showed that the rate of sodium azide decomposition was proportional to the surface area of the reactant crystallites. Reaction commenced at the edges of thin crystallites and proceeded inwards parallel to the azide ions (i.e., in [111] direction). Jacobs and Tariq Kureishy [65] showed that sublimation of the metallic (Na) product was suppressed in the presence of argon (500 Torr) and a reproducible kinetic behavior was observed. The α-time curve fitted the Avrami-Erofeev equation (n = 2) with Ea = 159 kJ mol^{-1} for the range 550 to 660 K. A mechanism involving two-dimensional growth of linear filaments of sodium within the intergranular material was proposed. After proceeding the reaction up to about 66%, loss of liquid metal from the decomposed crystal resulted in changing the rate law. It is suggested that incorporated Fe^{+3} positions in the reactant provide electron traps as an intentional impurity which are the potential nucleus-forming sites. Such impurities is also believed to participate in the nucleation process during decompositions of BaN_6 and PbN_6. In general, cation additives those have a greater radius than that of the host cation, produce strain in the structure which is identified as reason for the observed decrease in the induction period of decomposition. Torkar, Hertzog *et al.* [39, 40] observed an initial autocatalytic stage ($d\alpha/dt = k\alpha$) during nucleation on the (hk0) planes of the hexagonal reactant platelets which was accompanied by the removal of lattice defects. During the later decelerator process, reaction was proceeded by a diffusion-controlled mechanism along the preferred (001) planes. Sodium ions at the salt-metal interface lowered the Fermi level of the metal, thus aiding electron transfer. From dielectric measurements on samples of sodium azide which had been heated to the decomposition onset point, Ellis and Hall found three distinct processes during the early stages of the reaction ($\alpha < 0.02$). The 'dipole' structures are discussed with reference to the defects present in the solid. Preirradiation of NaN_3 with UV results in the development of centres of colloidal Na metal [66].

$$2NaN_3(s) \rightarrow 2Na(s) + 3\,N_2(g) \tag{8}$$

Silver Azide

The induction and acceleratory periods in the decomposition of silver azide is short, and reaction is predominantly deceleratory due to the rapid initial covering of crystallite surfaces with metallic product [67]. Gray and Waddington found that the reaction could be described by the contracting volume model with $E_a = 150$ kJ mol^{-1} between 515 and 555 K. AgN_3 melts at 523 K. Bartlett *et al.* [68] also confirmed the applicability of the contracting volume model for reaction with Ea = 130 kJ mol^{-1} in the range 463 to 523 K. McLaren and Rogers [69] identified the first step in AgN_3 decomposition produced free electrons and azide radicals (positive holes). They also discussed the mechanism in the context of the available spectroscopic and photochemical data. The rate of reaction is controlled by electron transfer and increased by addition of semiconducting oxides.

Incorporation of organic dyes also accelerate the decomposition rates and the longer dye chain is more favorable, probably due to the increased delocalization of the p-electrons. The catalytic effects result from the facile electron transfer in the reactant. Decomposition of silver azide in vacuum, induced by an electron beam, yields ultrafine particles of silver, predominantly less than 100 nm in diameter, and in the form of well-defined hexagonal or trigonal crystals and polygonized spheres. This technique of nanoparticle synthesis is recommended rather than the other methods [70].

$$2AgN_3(s) \rightarrow 2Ag(s) + 3N_2\ (g) \qquad (9)$$

Lead Azide

Because of the extensive technical use of lead azides as a primary explosives, there are lots of studies corresponding to their thermal stability. There are four polymorphic forms for lead azide in which the orthorhombic α-form is the most stable form. The other forms can be prepared under specific conditions. At high temperatures (above 613K), the reaction accelerates to detonation. This autocatalytic process has been attributed to an increased concentration of defect sites. According to the decomposition of α-lead azide, the α-time curves are predominantly deceleratory similar to those of silver salts. Griffiths and Groocock [71] studied this reaction using "well aged" α-lead azide samples of unusually homogeneous crystallite size distribution. There was an initial rapid evolution of gas upon heating ($E = 63$ kJ mol^{-1}) which was attributed to the decomposition of basic lead carbonate on crystallite surfaces. The acceleratory period of the main reaction was described by the power law ($n = 5$) with E value about 1100 kJ mol^{-1} and subsequently, the contracting volume equation. It is believed that nucleus formation continue during the acceleratory stage of reaction resulting in the interaction of positive holes to yield nitrogen. Growth is attributed to the reaction of positive holes with metal at the reaction interface. Therefore, the electron transfer step controls the salt decomposition. Some evidence for the existence of a reactant-product interface was obtained by microscopic examination.

Young [72] described a somewhat different pattern of kinetic behaviour for decomposition of the same salt. Again, there was an initial reaction of a surface impurity (about 2%) by a first-order process (E_a = 121 kJ mol^{-1}). The subsequent acceleratory reaction which was perhaps delayed by the initial process was fitted by the power law (n = 3). This is attributed to the growth of a constant number of nuclei present at time (t_0) and, because of the dependency of the number of such nuclei on temperature. The value of E_a for interface advance is 209 kJ mol^{-1}. The mechanism of reaction was not further developed.

Hutchinson *et al.* [73] have examined the effects of particle size on the decomposition of lead azide powder. Fine powders (radius = 8 nm) decomposed more than twice as rapidly as the large ones (radius = 24 nm). A mathematical model have been developed that incorporated the independently measured particle size distribution and gave an excellent description of the experimental sigmoid α-time curves. The influences of a variety of additives on decomposition of PbN_6 have been studied either incorporated in the crystal or in contact with the surfaces. Hutchinson and Stem [74] showed that incorporation of Fe^{2+} accelerates the reaction, but decreases the induction period, compared to the presence of $(FeN_3)^{2+}$ ion.

Savelev *et al.* recognized the bimolecular combination of positive holes as rate-limiting step. This conclusion arised from the observed influence of Cu^{2+} and Ag^{+} addition on reaction rates. Cu^{2+} occupies interstitial positions and aids migration of positive holes and Ag+ will be replaced with Pb^{2+} at lattice sites. Coating PbN_6 crystals with silver accelerated the rate of decomposition. Investigations have been made for influence of both low and high molecular mass organic dye molecules on reaction rate [75].

The possible formation of intermediates during the decomposition of lead azide has been discussed, but results were inconclusive. Shechkov suggested that up to 5% of the product nitrogen may be temporarily retained in the crystal and this may be expected to influence the kinetic behavior.

References

[1] Christian, I., *The Theory of Transformations in Metals and Alloys*, 816. 1965: Pergamon Press, Oxford.

[2] Zeyen, C., CNR Rao and KJ Rao: *Phase Transitions in Solids.* McGraw Hill Book Co. 1978. 330 Seiten, Preis: DM 81, 10. *Berichte der Bunsengesellschaft für physikalische Chemie* [*Reports of the Bunsen Society for Physical Chemistry*], 1979. 83(10): p. 1050-1050.

[3] Solymosi, F., *Structure and stability of salts of halogen oxyacids in the solid phase*. 1977: Wiley.

[4] Ubbelohde, A.R., *Melting and Crystal Structures*. 1965: Oxford University Press.

[5] Höhne, G.W.H., W. Hemminger, and H.-J. Flammersheim, Theoretical fundamentals of differential scanning calorimeters, in *Differential Scanning Calorimetry*. 1996, Springer. p. 21-40.

[6] Privalov, P.L. and S.A. Potekhin, [2] Scanning microcalorimetry in studying temperature-induced changes in proteins, in *Methods in Enzymology*. 1986, Academic Press. p. 4-51.

[7] Even, J., et al., Optical and calorimetric studies on the role of lattice mode softening in assisting a thermally enhanced solid state reaction. *Chemical Physics Letters*, 1997. 267(5-6): p. 585-589.

[8] Lin, L., et al., Calorimetric studies of the N-terminal half-molecule of transferrin and mutant forms modified near the Fe3+-binding site. *Biochemical Journal*, 1993. 293(2): p. 517-522.

[9] Protasevich, I., et al., Conformation and thermal denaturation of apocalmodulin: role of electrostatic mutations. *Biochemistry*, 1997. 36(8): p. 2017-2024.

[10] Von Stockar, U. and I.W. Marison, The use of calorimetry in biotechnology, in *Bioprocesses and Engineering*. 1989, Springer. p. 93-136.

[11] Weber, P.C. and F.R. Salemme, Applications of calorimetric methods to drug discovery and the study of protein interactions. *Current Opinion in Structural Biology*, 2003. 13(1): p. 115-121.

[12] Varghese, N., et al., A calorimetric investigation of the assembly of gold nanorods to form necklaces. *Chemical Physics Letters*, 2008. 450(4-6): p. 340-344.

[13] Haines, P., M. Reading, and F. Wilburn, Differential thermal analysis and differential scanning calorimetry, in *Handbook of thermal analysis and calorimetry*. 1998, Elsevier. p. 279-361.

[14] Danley, R.L., New heat flux DSC measurement technique. *Thermochimica acta*, 2002. 395(1-2): p. 201-208.

[15] Morgan, D., ME Brown. Introduction to Thermal Analysis: Techniques and Applications. London and New York (Chapman and Hall), 1988. viii+ 211 pp. *Mineralogical Magazine*, 1989. 53(373): p. 662-662.

[16] Pfeil, A. and S. Löbbecke, Controlled pyrolysis of the new energetic binder azide polyester PAP-G. *Propellants, explosives, pyrotechnics*, 1997. 22(3): p. 137-142.

[17] Smith, P.A.S., *The Chemistry of Open-chain Organic Nitrogen Compounds: Derivatives of oxidized nitrogen: hydrazines to nitrates*. Vol. 2. 1966: WA Benjamin.

[18] Bräse, S. and K. Banert, *Organic azides: syntheses and applications*. 2010: John Wiley & Sons.

[19] Scriven, E.F. and K. Turnbull, Azides: their preparation and synthetic uses. *Chemical Reviews*, 1988. 88(2): p. 297-368.

[20] Falvey, D.E. and A.D. Gudmundsdottir, *Nitrenes and Nitrenium Ions*. Vol. 11. 2013: John Wiley & Sons.

[21] Kolb, H.C., M. Finn, and K.B. Sharpless, Click chemistry: diverse chemical function from a few good reactions. *Angewandte Chemie International Edition*, 2001. 40(11): p. 2004-2021.

[22] Binder, W.H. and C. Kluger, Azide/alkyne-"click" reactions: applications in material science and organic synthesis. *Current Organic Chemistry*, 2006. 10(14): p. 1791-1815.

[23] Haiges, R., et al., Polyazide chemistry: preparation and characterization of Te (N3) 4 and [P (C6H5) 4] 2 [Te (N3) 6] and evidence for [N (CH3) 4][Te (N3) 5]. *Angewandte Chemie International Edition*, 2003. 42(47): p. 5847-5851.

[24] Agrawal, J.P. and R. Hodgson, *Organic chemistry of explosives*. 2007: John Wiley & Sons.

[25] Huynh, M.H.V., et al., Synthesis, characterization, and energetic properties of diazido heteroaromatic high-nitrogen C- N compound. *Journal of the American Chemical Society*, 2005. 127(36): p. 12537-12543.

[26] Abbenante, G., G.T. Le, and D.P. Fairlie, Unexpected photolytic decomposition of alkyl azides under mild conditions. *Chemical Communications*, 2007(43): p. 4501-4503.

[27] Harvey, G.R. and K. Ratts, Synthesis of azirenes from allenic esters. *The Journal of Organic Chemistry*, 1966. 31(12): p. 3907-3910.

[28] Boyer, J. and F. Canter, Alkyl and Aryl Azides. *Chemical Reviews*, 1954. 54(1): p. 1-57.

[29] Griess, J.P., XX. On a new class of compounds in which nitrogen is substituted for hydrogen. *Proceedings of the Royal Society of London*, 1864. 13: p. 375-384.

[30] Curtius, T., Ueber stickstoffwasserstoffsäure (azoimid) N3H [About nitrogenic acid (azoimide) N3H.]. *Berichte der deutschen chemischen Gesellschaft*, 1890. 23(2): p. 3023-3033.

[31] Lang, S. and J. Murphy, Azide rearrangements in electron-deficient systems. *Chemical Society Reviews*, 2006. 35(2): p. 146-156.

[32] Keicher, T. and S. Lobbecke, *Lab-scale synthesis of azido compounds: safety measures and analysis*. 2010: John Wiley and Sons: Chichester, UK.

[33] Meldal, M. and C.W. Tornøe, Cu-catalyzed azide- alkyne cycloaddition. *Chemical Reviews*, 2008. 108(8): p. 2952-3015.

[34] Katsuki, T., Azide compounds: Nitrogen sources for atom-efficient and ecologically benign nitrogen-atom-transfer reactions. *Chemistry Letters*, 2005. 34(10): p. 1304-1309.

[35] Bamford, A.F., M.D. Cook, and B.P. Roberts, Mechanism of the reaction of trialkylboranes with a-azidostyrene. *Tetrahedron Letters*, 1983. 24(35): p. 3779-3782.

[36] L'Abbe, G. and G. Mathys, Mechanism of the thermal decomposition of vinyl azides. *The Journal of Organic Chemistry*, 1974. 39(12): p. 1778-1780.

[37] Kedrowski, B.L., Synthesis of orthogonally protected (R)-and (S)-2-methylcysteine via an enzymatic desymmeterization and Curtius rearrangement. *The Journal of Organic Chemistry*, 2003. 68(13): p. 5403-5406.

[38] Shao, H., et al., A practical solid phase synthesis of quinazoline-2, 4-diones. *Tetrahedron Letters*, 1998. 39(40): p. 7235-7238.

[39] Wolff, O. and S.R. Waldvogel, Reliable protocol for the large scale synthesis of diphenylphosphoryl azide (DPPA). *Synthesis*, 2004. 2004(08): p. 1303-1305.

[40] Xiao, X., et al., Iridium (III)-Catalyzed Regioselective Intermolecular Unactivated Secondary Csp3- H Bond Amidation. *Angewandte Chemie International Edition*, 2016. 55(39): p. 11897-11901.

[41] Xu, H.D., et al., Copper-Catalyzed Cyclization/aza-Claisen Rearrangement Cascade Initiated by Ketenimine Formation: An Efficient Stereocontrolled Synthesis of a-Allyl Cyclic Amidines. *Angewandte Chemie International Edition*, 2014. 53(35): p. 9284-9288.

[42] Kuzmin, A., et al., Exploring photochemistry of p-bromophenylsulfonyl, p-tolylsulfonyl and methylsulfonyl azides by ultrafast UV-pump–IR-probe spectroscopy and computations. *Physical Chemistry Chemical Physics*, 2016. 18(12): p. 8662-8672.

[43] Lwowski, W. and E. Scheiffele, Curtius and Lossen Rearrangements. I. The Benzenesulfonyl System1. *Journal of the American Chemical Society*, 1965. 87(19): p. 4359-4365.

[44] Abramovitch, R.A., W.D. Holcomb, and S. Wake, The decomposition of. beta.-phenethylsulfonyl azides. Solution chemistry and flash vacuum pyrolysis. *Journal of the American Chemical Society*, 1981. 103(6): p. 1525-1533.

[45] Fair, H. and R. Walker, Energetic Manerials. vol. 1. *Physics and Chemistry of the Inorganic Azides.* New-York-London. Plenum Press., 1977: p. 382.
[46] Barbour, R.T., *Pyrotechnics in Industry.* 1981: McGraw-Hill Companies.
[47] ŠATAVA, V. and F. ŠKVÁRA, Mechanism and kinetics of the decomposition of solids by a thermogravimetric method. *Journal of the American Ceramic Society*, 1969. 52(11): p. 591-595.
[48] Bertrand, G., J.P. Majoral, and A. Baceiredo, Photochemical and thermal rearrangement of heavier main-group element azides. *Accounts of Chemical Research*, 1986. 19(1): p. 17-23.
[49] Gray, P., Chemistry of the inorganic azides. *Quarterly Reviews*, Chemical Society, 1963. 17(4): p. 441-473.
[50] Evans, B., A. Yoffe, and P. Gray, Physics and chemistry of the inorganic azides. *Chemical Reviews*, 1959. 59(4): p. 515-568.
[51] Richter, T., *Energetic Materials 1: Physics and Chemistry of Inorganic Azides.* Plenum, New York, 1977: p. 33.
[52] Galwey, A.K., R. Spinicci, and G.G. Guarini, Nucleation and growth processes occurring during the dehydration of certain alums: the generation, the development and the function of the reaction interface. *Proc. R. Soc. Lond. A*, 1981. 378(1775): p. 477-505.
[53] Wischin, A., The thermal decomposition of crystals of barium azide. *Proc. R. Soc. Lond. A*, 1939. 172(950): p. 314-325.
[54] Mott, N.F., On the decomposition of metallic azides. *Proc. R. Soc. Lond. A*, 1939. 172(950): p. 325-335.
[55] Thomas, J. and F.C. Tompkins, The photodecomposition of barium azide. *Proc. R. Soc. Lond. A*, 1951. 209(1099): p. 550-561.
[56] Torkar, K. and H. Spath, Die thermische Zersetzung von Bariumazid-Einkristallen, 1. Mitt.: Die formale Kinetik des Zerfalls von Ba (N 3) 2-Einkristallen [The thermal decomposition of barium azide single crystals, 1st center: The formal kinetics of the decay of Ba (N 3) 2 single crystals]. *Monatshefte für Chemie und verwandte Teile anderer Wissenschaften*, 1967. 98(5): p. 1696-1711.

[57] Salot, S. and J.C. Warf, Barium pernitride from thermal decomposition of barium azide. *Inorganic Chemistry*, 1974. 13(7): p. 1776-1778.

[58] Prout, E. and D. Moore, The thermal stability of ?-irradiated strontium azide. *Journal of Inorganic and Nuclear Chemistry*, 1969. 31(6): p. 1595-1607.

[59] Prout, E. and E. Shephard, The coirradiated decomposition of barium and strontium azides. *Journal of Inorganic and Nuclear Chemistry*, 1981. 43(9): p. 1977-1980.

[60] Tompkins, F. and D. Young, Decomposition of calcium azide. *Transactions of the Faraday Society*, 1965. 61: p. 1470-1480.

[61] Prout, E. and V. Liddiard, Thermal decomposition of lithium azide. *The Journal of Physical Chemistry*, 1968. 72(6): p. 2281-2283.

[62] Haase, J., Large-scale preparation and usage of azides. *Organic Azides: Syntheses and Applications*, 2010: p. 29-51.

[63] Betterton, E.A., Environmental fate of sodium azide derived from automobile airbags. *Critical reviews in Environmental Science and Technology*, 2003. 33(4): p. 423-458.

[64] Walker, R., N. Gane, and F.P. Bowden, The influence of crystal size and crystallographic orientation on decomposition in the solid state: sodium and thallous azides. *Proc. R. Soc. Lond. A*, 1966. 294(1439): p. 417-436.

[65] Jacobs, P. and A.T. Kureishy, 910. Kinetics of the thermal decomposition of sodium azide. *Journal of the Chemical Society (Resumed)*, 1964: p. 4718-4723.

[66] Ellis, P. and P. Hall, Dielectric investigation of the thermal decomposition of sodium azide. *Transactions of the Faraday Society*, 1968. 64: p. 1034-1040.

[67] Garner, W.E., *Chemistry of Solid State*. Butterworths, 1955.

[68] Bartlett, B., F.C. Tompkins, and D. Young, The decomposition of silver azide. *Proc. R. Soc. Lond. A*, 1958. 246(1245): p. 206-216.

[69] McLaren, A. and G. Rogers, The optical and electrical properties of silver azide and their relation to its decomposition. *Proc. R. Soc. Lond. A*, 1957. 240(1223): p. 484-498.

[70] Herley, P.J. and W. Jones, Electron microscopic study of ultra-fine particles of silver resulting from the electron-beam-induced decomposition of silver azide. *Journal of the Chemical Society*, Faraday Transactions, 1992. 88(21): p. 3213-3216.

[71] Groocock, J., The effect of high energy X-rays and pile radiation on the thermal decomposition and thermal explosion of a-lead azide. *Proc. R. Soc. Lond. A*, 1958. 246(1245): p. 225-232.

[72] Young, D., Thermal Decomposition Of Alpha-Lead Azide. *Journal of the Chemical Society*, 1964(SEP): p. 3141-&.

[73] Hutchinson, R.W., S. Kleinberg, and F.P. Stein, Effect of particle-size distribution on the thermal decomposition of. alpha.-lead azide. *The Journal of Physical Chemistry*, 1973. 77(7): p. 870-875.

[74] Hutchinson, R.W. and F.P. Stein, Effect of iron impurities on the thermal decomposition of. alpha.-lead azide. *The Journal of Physical Chemistry*, 1974. 78(5): p. 478-481.

[75] Haberman, J. and T.C. Castorina, Effect of adsorbates on time of thermal initiation of lead azide. *Thermochimica Acta*, 1972. 5(2): p. 153-164.

In: Thermal Decomposition
Editor: Kalyna Jocosa Pace
ISBN: 978-1-53615-460-3

Chapter 2

CATALYTIC THERMAL DECOMPOSITION AND COMBUSTION OF HTPB-AP COMPOSITE SOLID PROPELLANTS IN PRESENCE OF NANOSIZED BIMETALLIC TMOS

Mohan Varma**[*] **and N. Lokeshwaran
Department of Space Engineering and Rocketry
Birla Institute of Technology, Mesra, Ranchi, India

ABSTRACT

Thermal decomposition studies have been carried out on nano-catalyzed compositions of AP, HTPB and its propellants and compared with that of pure AP, HTPB and uncatalyzed HTPB-AP propellants. Iron oxide and copper chromite/oxide have been extensively used as burn rate modifiers for composite solid propellants. This study aims to assess the

[*] Corresponding Author's E-mail: mohanvarma@bitmesra.ac.in.

catalytic efficacy of nanosized bimetallic transition metal oxides of copper - iron and cobalt - iron synthesized using the solution combustion method in a laboratory reactor. The prepared catalysts were characterized for their structure and particle size. It has been observed that addition of 3 wt. % $CoFe_2O_4$ and $CuFe_2O_4$ to AP, significantly lowers its high temperature decomposition temperature by 44.6°C and 79.3°C respectively. However, these catalysts have shown negligible influence on thermal decomposition behavior of fuel binder alone. Burn rate studies carried out in a Crawford high pressure strand burner to assess the variation of burn rate with pressure for virgin and catalyzed compositions showed that the addition of 3 wt.% of $CoFe_2O_4$ gave the highest burn rate of 16.97 mm/s at 6.205 MPa. Burn rate studies of propellants containing 3 wt. % of Fe_2O_3 and CuO were also carried out to compare the burn rate characteristics of synthesized catalysts. It can be concluded that $CoFe_2O_4$ appears to be a better burn rate enhancer than commercial Fe_2O_3 which is generally used in composite solid propellant formulations. The combustion index, however, is found to increase with the increase in the concentration of $CoFe_2O_4$ in propellant composition. The composition giving the highest burn rate also exhibits a higher value of combustion index of the order of 0.66. Addition of aluminum powder along with meagre amounts of activated charcoal and $CoFe_2O_4$ in a formulation is seen to give a 30% increase in burn rate of the base composition at 80 wt. % solid filler loading. This study establishes that the synthesized bimetallic nano-sized TMOs can be superior burn rate modifiers in comparison to conventional TMOs.

Keywords: solid rocket motor, composite solid propellants, thermal decomposition, burn rate, nano-catalyst

INTRODUCTION

Solid rocket motors extensively use composite solid propellants (CSP) that are normally composed of a polymeric fuel binder, inorganic crystalline oxidizer, plasticizer and curing/crosslinking/bonding agents to provide required ballistic and mechanical properties. One of the most important performance parameters of these heterogenous propellants is the burning rate that involves the steady linear regression of the propellant burning surface in a direction essentially perpendicular to the propellant surface. Rocket motor design requires in depth understanding of burn rate behavior of the chosen propellant under broad motor operation and design limit conditions [1]. Solid rocket motor, once ignited, stops only after consumption of total propellant mass cast in its combustion chamber. The combustion process cannot be terminated during operation. As a result, burning time for a rocket engine and the chamber pressure or thrust produced during operation reflect its performance. The thrust-time profile is often used to decide application areas and acceptability for a mission. The net burning rate is the result of condensed phase reactions in solid state beneath the burning surface that are dependent on heat transfer, surface reactions initiated due to thermal decomposition of oxidizer and initial breakdown of the polymeric binder and finally the gas phase reactions that involve complex combustion processes in the flame zone. A wide range of physicochemical parameters that may influence reaction mechanisms in each stage of combustion through heat transfer, fluid (combustion gases) flow and variation in operating conditions affect the propellant burn rate significantly. The major contributing factors in influencing burn rate of solid propellant may be chamber pressure, initial propellant temperature, oxidizer particle size used in the propellant formulation, oxidizer-fuel mixture ratio, overall chemical formulation of the propellant, transverse flow velocity of combustion gases across burning surface and nature of propellant grain shape factor. The central perforation in propellant grain is provided to achieve a desired thrust variation with time during rocket motor operation.

There are several methods [2] that are used in solid rocket industry to measure the burn rate of propellants at laboratory scale and under conditions that simulate actual motor operating conditions at sub-scale and similar to ballistic evaluation motor test. The laboratory scale methods generally use a few centimeter-long propellant strands that is ignited using an electrical coil at one end and thin fuse wires are embedded in strand at two locations to signal start and stop stage on arrival of flame to a connected electronic timer. Crawford bomb [3-5], close bomb technique [6], microwave interferometry method [7], ultrasonic method [8, 9], real time x-ray radiography [10, 11], optical techniques [12, 13], laser technique [14, 15] or acoustic emission technique [16] may be used to determine the burn rate of CSP. Other methods employ subscale motor firing below a thrust level of 400 N [17, 18] or ballistic evaluation motor test [19] that are modified from time to time with development of better transducers, sensors and high-speed recording devices.

Use of solid propellants dates to fire projectiles of different caliber used in 13th century by Chinese, Indians, Mongols and Persians in its nascent form. Several advancements took place in the interim period and double base solid propellant containing nitrocellulose and nitroglycerine were developed but these had safety, performance and ageing characteristics related issues. The first castable composite solid propellant of asphalt as binder and oxidant potassium perchlorate was developed at Caltech, USA in 1942. Later, substituted elastomeric rubber for asphalt was used to develop a flexible and geometrically stable solid propellant shape that could be cast inside a solid rocket motor casing at the same laboratory. The developed composite solid propellants in subsequent years had to address issues related to improved ballistic properties and high performance in terms of specific impulse, long shelf-life as these were to be stored in missiles in battle ready condition, excellent mechanical properties to sustain high acceleration during flight and high solid filler (oxidizer and metal powder) loading in binder to improve energetics. The progress in development of composite solid propellants after second world war has been phenomenal and today, we have solid rocket motors that deliver precise thrust level of few Newtons in micro thrusters to nearly 14000 kN for space shuttle boosters. Similarly, the propellant weight in propulsion devices may also vary from few grams for micro-thruster to nearly 600 tons for largest booster stage.

Modern composite solid propellants comprise of a fuel binder like carboxy terminated polybutadiene (CTPB) or hydroxy terminated polybutadiene (HTPB) or polybutadiene acrylonitrile (PBAN) ($\leq$ 14 wt.%) and ammonium perchlorate (AP $\leq$ 86 wt.%) has emerged as work horse oxidizer in most of the present-day formulations. Aluminum metal powder of different particle sizes are also used as a constituent between 16 to 18 wt.% by replacing AP to enhance energetics of composite solid propellants. AP containing propellants, however, emit HCl gas on combustion that poses very serious problem when tons of propellant is burnt in a large booster motor. Alternate chlorine free oxidizers like ammonium dinitramide (ADN) and hydrazinium nitroformate (HNF) are also being tried in new formulations along with high energy binders such as glycidyl azide polymer (GAP), poly dicyclopentadiene (PDCPD) and bis azido methyl oxetane (BAMO) and its copolymers [20]. However, new fuel binder systems are yet to be used in space missions where solid propellant is to be used in large quantity.

Application of composite solid propellants in rocket motors varies from laboratory experiments related to their development and testing and final utilization in sounding rockets, launch vehicles and tactical missiles. Other civilian application of solid propellants includes safety devices with special purposes in automotive and aviation industries such as inflators in air bags. These are also used in pyrotechnic valves, actuating devices, emergency separation systems, gas generators and devices designed to operate auxiliary propulsion systems. These variations in usage demand tailoring of burn rate, flame temperature, energetics and required combustion index that is determined by evaluating pressure response of burn rate in a broad range of operating conditions and it has immense influence on rocket motor ballistics. The changes have traditionally been brought about by addition of catalyst or additive like ferric oxide [21-25], copper chromite [22, 26, 27], ferrocene and its derivatives [28-30], lithium fluoride [31] or in recent times by nano-transition metal oxides (TMO) and similar materials [32-36].

COMBUSTION MECHANISM

Combustion mechanism of composite solid propellants is very complex in nature and involves intricacies of constituting materials, chemical kinetics, thermodynamics, heat transfer, flow of combustion gases across burning surface individually and mutual interaction of these in pressure dependent combustion process of solid propellants. Several attempts have been made in the past to understand the details of combustion phenomenon and combustion models have been proposed to explain and predict burning rate of CSP in ambient and operating conditions including role of various factors that influence it. The first comprehensive review [37] on the subject is available in literature. The approach of building elements of combustion model and related assumptions gave highest importance to combustion of AP that may constitute up to 86 wt.% of the total mass of CSP and both condensed phase theories [38] and gas phase theories [39] of AP combustion form the backbone of subsequent developments in combustion modeling of CSP. The sublimation, thermal decomposition and deflagration behavior of AP and its properties like enthalpy, activation energy, stoichiometry of decomposition and effect of pressure and temperature on thermal decomposition of AP has been the subject of many investigations that are summarized by Kishore and Sridhara [40]. In comparison, there has been relatively less attention given to degradation and pyrolysis of fuel binders that are normally used in CSP formulations primarily with argument that its concentration remains much lower in comparison to the oxidizer and so its role in the combustion process shall be of limited importance. However, intensive investigations are needed to understand the pyrolysis of polymeric fuel individually and in propellant combustion under different operating conditions. An earlier review on the subject [41] nicely brings out the essential features of pyrolysis of various fuel binders. The mechanism of pyrolysis of fuel binder generally involves a linear process at large heat flux that leads to generation of degradation products at its surface. The kinetics of polymeric fuel pyrolysis is understood to be independent of pressure, presence of the oxidizer or catalyst and is associated with a molten layer mingled with char. There is still difference of opinion whether the surface temperature of binder remains the same as that of oxidizer (AP) during combustion of CSP. Various combustion models proposed for CSP assign much greater importance to involvement of fuel vapors in gas phase in comparison to surface regression phenomenon that results due to quasi-one-dimensional heat transfer from propellant flame back to the solid surface. A typical temperature profile of burning propellant in different zones is presented in Figure 1.

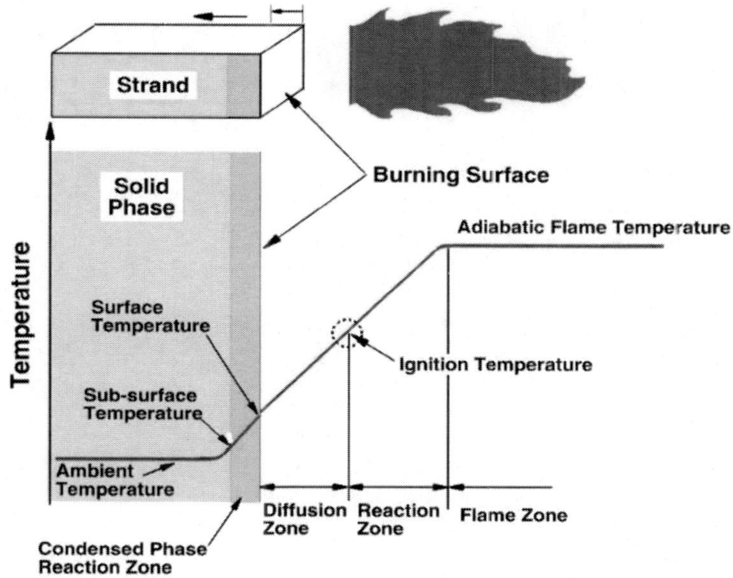

Figure 1. Combustion wave structure of a typical CSP.

Above the burning surface of composite solid propellant, there exists a diffusion zone where gaseous products produced by the decomposition of oxidizer and pyrolysis of fuel binder intermix and exothermic reactions take place in reaction zone that go on to achieve adiabatic flame temperature. The heat transferred from this zone to the sub-surface of the burning propellant trigger condensed phase reactions raising temperature beneath the burning surface. Importance of solid-phase reactions is well known, and these have significant influence on burning rate of CSP. It is believed that sub-surface gets heated at a rate of about 10°C/minute as CSPs normally show poor thermal conductivity. Under normal conditions, the flame is attached to burning surface in case of AP based CSPs, however, at sub-ambient pressures there may be a visible stand-off distance between solid surface and the flame due to reduced chemical activity in combustion process. A general description of important combustion models [42-45] that aim to explain the combustion of CSP with associated flames and the role of different factors influencing burn rate, is briefly discussed. The models primarily aim to predict the burn rate under conceived conditions and their results are validated experimentally.

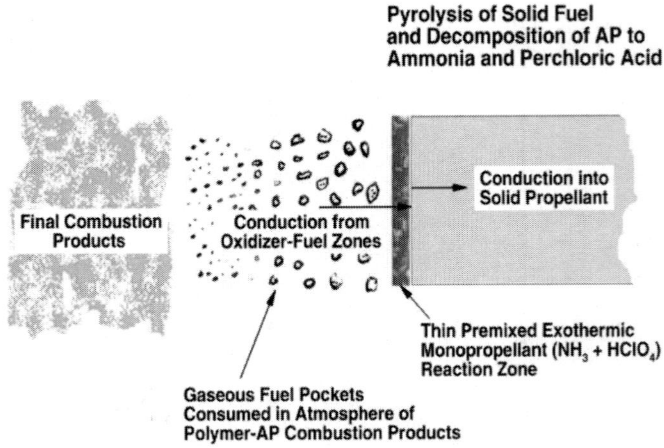

Figure 2. GDF model of AP based composite propellant.

The first noteworthy attempt towards explaining the burning mechanism of AP based propellants was made by Summerfield [42] and this approach is known as Granular Diffusion Flame (GDF) model as depicted in Figure 2.

The model assumed that a thin film (about 500 μm) covers the dry propellant burning surface. The oxidizer rich decomposition products of AP which are produced as ammonia and perchloric acid in a premixed reaction zone diffuse into the fuel rich gases present as distinct pockets of finite mass content, released from fuel binder. These pockets burn in oxidizing vapors whose concentration depends on AP particle size. The diffusion flame thus produced transfers heat back to surface for sustained gasification of polymer and AP. Further, it is considered that this diffusion process is not turbulent, heat transfer occurs only by conduction and reaction kinetics follows well known Arrhenius law. The final diffusion flame height from surface is seen to be inversely proportional to pressure to the power 1/3. At lower pressure, molecular diffusion is predominant to chemical reaction and reverse is true at higher pressures, however, at intermediate pressures combustion is controlled by a mixed response of both the factors. The validation of model used experimental results obtained with AP-styrene polyester resin CSP using a bimodal AP. The inherent shortcomings of this model include the experimental determination of constants of pressure dependence of burn rate, no weightage to effect of fuel binder used in formulation and weak influence of particle size of AP is reflected. The model nicely predicts the effect of copper chromite as catalyst on burn rate. The GDF model has the advantage of simplicity and easy validation for several types of CSPs.

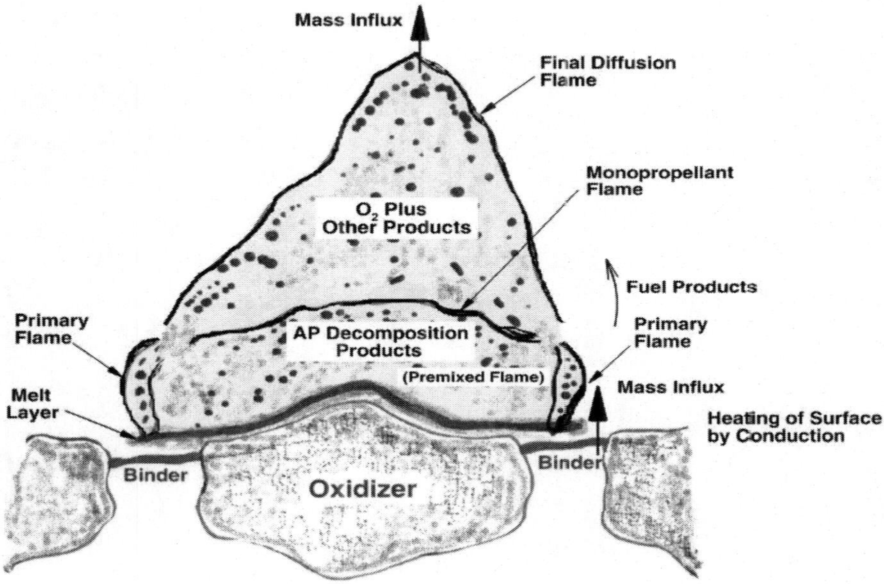

Figure 3. BDP model of multiple flames for CSP.

Hermance [43] provided the extension of GDF model and addressed several important aspects like prediction of effect of fuel binder, very weak influence of AP particle size on burn rate at high pressures (above 27 MPa). The model included the heterogeneity of burning surface in composite propellants into mathematical formulation. The essential features included endothermic pyrolysis of binder, exothermic decomposition of oxidizer, exothermic chemical reaction between binder and oxidizer and gas phase combustion of products of decomposition of fuel and oxidizer. The interfacial reactions between fuel binder and oxidizer below surface temperature (between 400°C – 700°C) has also been considered. The key features in this approach include mass fluxes released from fuel and oxidizer and mean planar fuel surface is assumed to intersect the oxidizer particle whose diameter goes on decreasing with reaction time, however, a customary single average particle size distribution is replaced by 10% and 90% points of distribution. The model provides a good fit of burn rate in a pressure range from ambient to 40 MPa and accurately predicts the effect of oxidizer particle size on burn rate of CSP at higher pressures (≥20 MPa).

Another important development in conceptualization of combustion of CSP was proposed in BDP model [44] that is popularly known as 'multiple flames model'. It addressed the complexity of the real combustion phenomenon considering different types of flames, as observed in heterogenous propellants due to varied thermal response of its constituents. The schematic representation of the model is presented in Figure 3.

The model conceives AP monopropellant flame above the propellant surface that emanates due to decomposition of oxidizer into ammonia and perchloric acid (redox flame) and subsequent diffusion flame may be located below this flame where binder has begun to pyrolyze due to heat transferred back to the surface. Final diffusion flame above the AP flame is a result of abundant fuel vapors and secondary oxidizing products that are left unconsumed in AP flame. Further, at low pressure between ambient to about 0.7 MPa, chemical reactions seem to be the rate controlling process as complete diffusion of fuel and oxidizer vapors produce the primary diffusion flame. The existence of possible melt layer above solid surface also depends on propellant composition and operating pressure and after a critical pressure level it may disappear. At relatively higher pressures, the diffusion process is inhibited and affects the AP flame at center of oxidizer surface. The diffusion of fuel vapors into gaseous decomposition products of AP becomes increasingly difficult at very high pressures and as a result primary diffusion flame becomes shorter and negligible fraction of fuel vapor can react in the formation of primary diffusion flame. Another model [45] that replaces one-dimensional flames of BDP model (considered for ease of calculation of diffusion rate, chemical kinetics and heat transfer) to actual unit flames observed in propellant combustion. This 'Petite Ensemble Model' considers planar and total surface areas of imbedded oxidizer particles burning in a circular duct that is surrounded by evolving concentric annulus streams of fuel. The associated mass flux is calculated for monodispersed subareas of each particle size that is used to arrive at burning rate of actual poly-dispersed propellant. The related mathematical treatment of models is beyond the scope of this chapter.

The models discussed above have some common features that are based on typical nature of propellant ingredients involved in CSP but influence of additional factors like types of flames and influence of pressure dependent chemical reactions, reaction zone thickness, standoff distance of flames from propellant surface and heat transfer etc. is needed to be clearly understood to predict the burn rate of CSP accurately under different operating conditions. A comprehensive model shall reduce the dependence on experimental investigations time and again and thus save money and time. In recent past, several attempts [46-50] have been made to address the limitations of older models due to our current ability to address details of intricate combustion processes (rather than taking a mean parametric value, global kinetics approach or unrealistic assumptions) and utilizing phenomenal advancement in computation techniques and machines. The elaborate understanding of the combustion phenomenon of virgin CSP would assist in choosing the appropriate catalyst and metal to improve their energetics and can further help in knowing the intricate mechanism of catalytic combustion of such systems.

CATALYTIC COMBUSTION

The need for altering burn rate of an established CSP formulation becomes necessary as per specific mission requirements or application. The suitably tailored burn rate can easily be achieved by incorporation of additives/catalyst in propellant formulation whose type, concentration and particle size can be used to manipulate the property to a higher or lower level, as desired. Generally, most missions require a faster burn rate, and catalysts have been traditionally selected on their ability to enhance oxidizer decomposition rate as it exceeds 80% of the propellant weight. As a result, several TMOs have been investigated for their effectiveness as catalyst for decomposition and deflagration of AP and subsequently ferrocene and its derivatives were also evaluated. TMOs of iron, nickel, cobalt and manganese are known to decrease the rate of polymer combustion as these oxides suppress the gasification process and subsequent gas phase reactions [51]. Little information is available in literature about exact mechanism of catalytic action on combustion of CSP in conditions encountered in combustion chamber of solid rocket motor. Most studies on catalytic efficacy are conducted under conditions encountered in condensed phase combustion. The enhanced decomposition of AP in presence of TMOs is attributed to their variable valence or oxidation states and capability to act as electron donor or acceptor. Electron or proton transfer mechanisms have been proposed to explain catalytic action by many researchers. A comprehensive review on the subject is available in literature [52].

The catalyzed decomposition and pyrolysis of fuel binders is much less attempted as it was felt that its presence in CSP remains normally less than 12-15 wt.% and it will have limited impact on overall burn rate of CSP. The catalysts in CSP formulations are normally added below 3 wt.% of the propellant mass as higher solid loadings may adversely affect the processing of propellant slurry and mechanical properties of the finished propellant. Unfortunately, the combustion behavior of resins and butadiene polymers used as fuel binder in CSP remains unaffected or retarded at low concentrations of additives used so far. Very high burn rate of solid propellants is required for their application in tactical missiles that require high acceleration. The efforts should now be directed to discover catalysts that not only accelerate the deflagration process of AP but also the pyrolysis rate of polymeric binder like HTPB. The fuel binder decomposition occurs at much higher temperature (478°C) than AP (295°C) [25]. Since, AP crystals are trapped in the polymeric matrix in a CSP (Figure 4), an earlier shift in decomposition temperature of binder in presence of suitable catalyst shall facilitate release of oxidizer particles from surface and simultaneously advance the gasification process of binder and its mixing with already decomposed oxidizer leading to enhanced vapor phase reactions and burn rate.

Figure 4. SEM photograph of HTPB-AP propellant.

The imbedded solid catalyst at low concentrations in a CSP are seen to cling on the surface of the solid oxidizer or move to the interparticle space between the oxidizer particles being much smaller in size as shown in Figure 5.

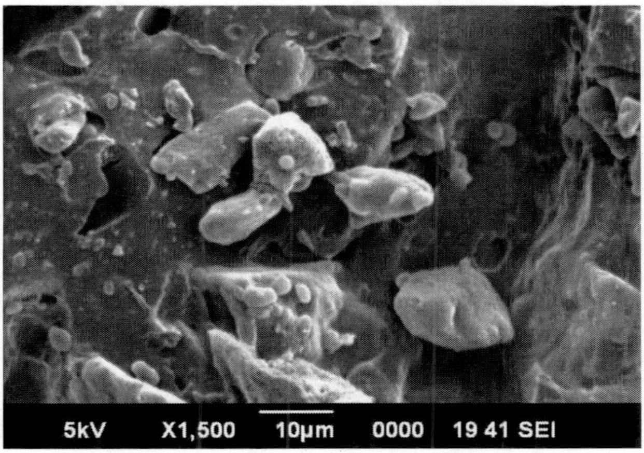

Figure 5. SEM of CSP with 2 wt.% of catalyst (2-3 μm).

The scanning electron microscope (SEM) picture also clearly shows tiny catalyst particles imbedded in binder matrix that may eventually influence the binder decomposition. A catalyst that enhances oxidizer decomposition process, but retards pyrolysis rate of binder may eventually improve the burn rate of CSP to a limited extent. As a result, future research on catalytic combustion of CSP shall be centered around search of complex nanosized TMOs that can prove much better than already used single metal TMO of micron size and/or use of another catalyst used specifically for enhancement of binder pyrolysis. This dual catalyst approach in combustion of CSP may promise to produce super-fast burn rates in solid propellants, in future. We report a promising new approach that uses complexing of metallic TMOs and their size effect in catalytic combustion of HTPB-AP based CSP.

EXPERIMENTAL WORK

The main objective of this work is to synthesize complex catalysts consisting of two different transition metals and to test their efficacy as a burn rate modifier for the HTPB-AP composite solid propellant. As mentioned earlier, nanosized transition metal oxides have proven to be superior burn rate modifiers in comparison to their micron sized counterparts. Therefore, the use of nanosized additives in CSP formulations has been of great interest in recent years. Attempt has been made to synthesize two nanosized crystalline complex metal oxides namely, Cobalt Iron oxide – $CoFe_2O_4$ and Copper Iron oxide – $CuFe_2O_4$ for its assessment as effective catalysts.

Preparation of Catalysts

Solution Combustion Synthesis (SCS) approach has been used to synthesize crystalline nanosized metal oxides for this study. Nano-crystalline metal oxides were prepared by rapidly heating aqueous solutions containing stoichiometric amounts of the respective metal nitrates—Cobalt (II) nitrate hexahydrate, Copper (II) nitrate trihydrate and Iron (III) nitrate nonahydrate (CDH, India) that act as oxidizer and Glycine (BDH) is taken as fuel. The surface area of the final product and its crystalline structure could be controlled by varying the experimental conditions such as the amount of water, initial temperature etc. Stoichiometric amounts of reactants were weighed and dissolved in distilled water. The resultant solution is used as precursor for solution combustion reaction and heated in a muffle furnace. The solution starts boiling and temperature of solution remains at 100°C ± 5°C until majority of the water evaporates and a gel is formed. The temperature then further rises slowly until the reactants start smoldering and product gases are released. Finally, the gel gets ignited when the temperature reaches the ignition point and a flame appears. This combustion wave travels through the reactants until all reactants are combusted resulting in a voluminous powdery product at preset temperature of 400°C. After this reaction is complete, the furnace is switched off and cooled to 25°C.

Characterization of Catalysts

It is important to characterize the synthesized catalysts to confirm their composition, surface morphology and to determine the crystal size and structure. X-ray diffraction (XRD) [Rigaku Smart lab X-ray Diffractometer] patterns were acquired at wavelength of 1.545 A° and the target used was pure metal target – Cu in values of 2θ ranging from 10° to 80°. Scanning Electron Microscopy (SEM) [JEOL JSM 6390 LV] was done in low vacuum and the acceleration voltage was kept at 10 kV. Images were taken up to a magnification of 2000 x in order to examine the pores in the powder, if any. Fourier transform infrared spectroscopy (FTIR) technique was used to obtain absorption or emission peaks over a range of wavelengths to characterize synthesized catalysts.

Formulation and Processing of CSP

The hydroxyl terminated polybutadiene (HTPB) [Number average molecular weight = 2300 and hydroxyl value of 41.2 mg KOH/g] used as fuel binder, was obtained from VSSC, Thiruvananthapuram, India and the oxidizer AP was procured from Pandian Chemicals Ltd, Madurai, India. The oxidizer was pulverized and sieved to get particle size of 250 µm as coarse and 75 µm for fine solid filler loading and solid filler loading was maintained at 82 wt.% in all propellant compositions. Dioctyl adipate (DOA) [Make: Sigma-Aldrich] was used as plasticizer and Toluene diisocyanate (TDI) [Merck] as curing agent and glycerol [Merck] as crosslinking agent were added constituents in the formulations. The synthesized catalysts - $CoFe_2O_4$ (CoIO), and $CuFe_2O_4$ (CuIO) were incorporated in formulations at a level of 1, 2 and 3 wt.% individually and samples containing 3 wt.% commercial Fe_2O_3 and CuO catalysts were also prepared for comparison purpose. The basic propellant composition is given in Table 1 and all other catalyzed compositions prepared in Table 2.

Table 1. Basic propellant composition

Propellant ingredient	HTPB (wt. %)	AP/solid filler (wt. %)	DOA (wt. %)	TDI (wt. %)	Glycerol (wt. %)
	13	82	3.73	1.12	0.15

Prior to processing of propellant, AP was dried in the oven to remove moisture. The propellant ingredients were then weighed in requisite quantity. The DOA was dehumidified and after cooling to room temperature mixed with HTPB in a mixer vessel. AP was then added in 3-4 instalments and mixing was done for an hour at a temperature of 60 ± 2°C. In case of catalyzed propellant formulations, the catalyst was mixed after first instalment of AP was added to ensure homogeneity and crosslinking agent, TDI was mixed in the slurry at the end and mixing was continued for another 15 minutes. Mixing at elevated temperature helped in achieving higher solid filler loadings and to maintain the flowable slurry viscosity. The propellant slurry was then cast into a rectangular pre-lubricated mold under vacuum (50 Torr) for 30 minutes to remove any air trapped in the slurry. Finally, the molds were kept in a hot air oven at 60 ± 2°C for six days to get a cured CSP for further investigations.

Table 2. Composition of control and catalyzed propellant formulations

S. No.	AP (wt. %)	Fuel binder (wt. %)	CoIO (wt. %)	CuIO (wt. %)	IO (wt. %)	CuO (wt. %)
1	82	18	-	-	-	-
2	81	18	1	-	-	-
3	80	18	2	-	-	-
4	79	18	3	-	-	-
5	81	18	-	1	-	-
6	80	18	-	2	-	-
7	79	18	-	3	-	-
8	79	18	-	-	3	-
9	79	18	-	-	-	3
10	100	-	-	-	-	-
11	99	-	1	-	-	-
12	98	-	2	-	-	-
13	97	-	3	-	-	-
14	99	-	-	1	-	-
15	98	-	-	2	-	-
16	97	-	-	3	-	-
17	97	-	-	-	3	-
18	97	-	-	-	-	3
19	-	100	-	-	-	-
20	-	97	3	-	-	-
21	-	97	-	3	-	-

Thermal Analysis and Burn Rate Studies

The thermal analysis studies to investigate effect of mixed TMOs and conventional TMOs on CSP and its ingredients were conducted on a Simultaneous Thermal Analyzer (STA) [Netzsch STA 409/PG] using an alumina crucible that contained 3-5 mg of sample. The experiments were conducted in a temperature range of 30°C to 700°C at a heating rate of 10°C/min under dry nitrogen environment, maintaining a flow rate of 25 ml/min.

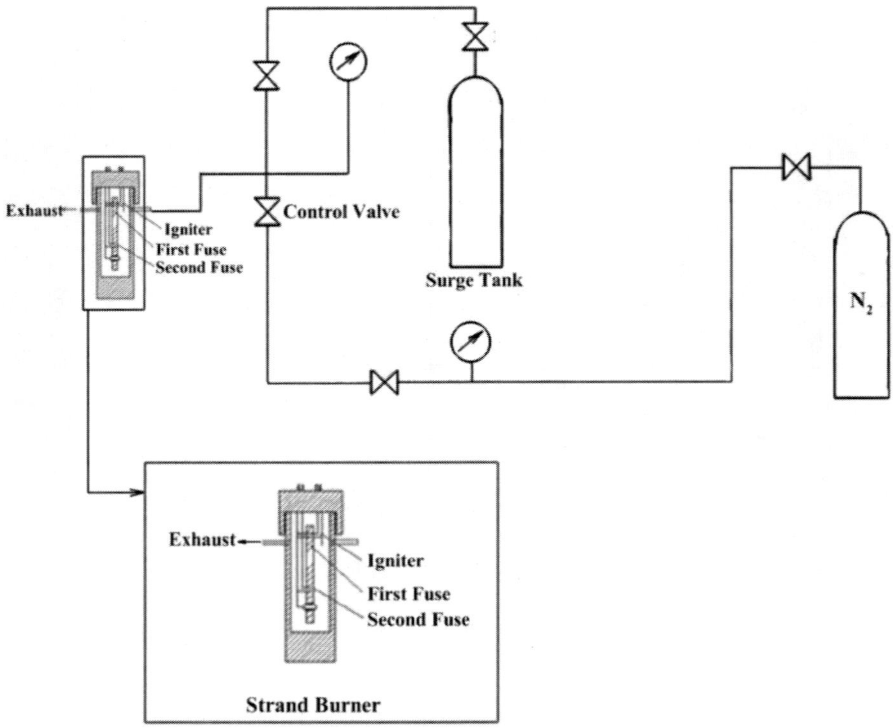

Figure 6. High pressure strand burner.

Burn rate studies on prepared CSP samples were carried out at three different chamber pressures (2.07, 4.14 and 6.21 MPa) in a high-pressure strand burner setup shown in Figure 6. Each sample strand was marked 6 cm apart and two fuse wires were inserted through a fine suture made in it. The first fuse wire was kept 1 cm below the igniter fire as shown in the figure. The strand burner bomb was then assembled and pressurized to a desired level. The strand was ignited in each case by passing electrical current through igniter wire. Once the flame reaches the first fuse the timer in the circuit starts counting and stops when the flame melts the second fuse wire. The time recorded for burning the given length of strand is used to calculate the linear burning rate of the CSP sample used.

RESULTS AND DISCUSSION

Characterization of Catalysts

X-Ray Diffraction Study

The catalysts were characterized by XRD in order to determine its average crystallite size and to confirm the molecular formula of the product. XRD was performed in the 2θ range of 10° - 80° for prepared catalysts using a Cu-Kα radiation with a wavelength of 1.54 A°. The XRD patterns for both samples were obtained from PDXL software provided by the manufacturer (Rigaku). The patterns obtained from the PDXL were scrutinized through 'Match' software to compare obtained peaks with the standard data Crystallography Open Database (COD) preloaded in software. The result for synthesized $CoFe_2O_4$ (CoIO) catalyst is shown in Figure 7.

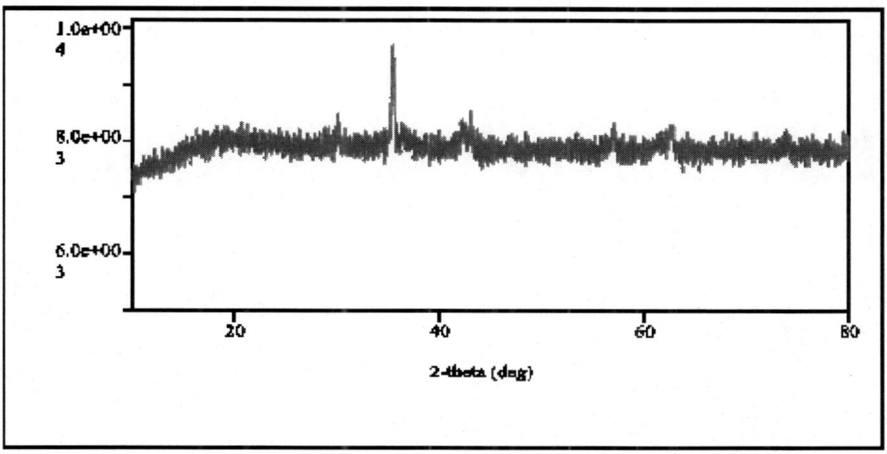

Figure 7. XRD pattern of CoIO.

Six distinct peaks were obtained at 30.17°, 35.571°, 43.23°, 57.13° and 62.75° for this catalyst. According to COD in 'Match' software, the molecular formula for these signature peaks was found to be $CoFe_2O_4$. The crystals are also found to have a spinel cubic structure which was also obtained from the 'Match' software by profile fitting the peaks with the pattern provided by COD. The crystallite sizes were obtained from the peak list data sheet generated by the PDXL software. The crystallite size of CoIO was found to be 44.66 nm from the XRD data which appears to be a good size for catalyst requirement.

The XRD for bimetallic TMO, $CuFe_2O_4$ (CuIO) is presented in Figure 8. The obtained pattern shows four distinct peaks at 30.32°, 33.24°, 35.71° and 43.37° which is characteristic of $CuFe_2O_4$ spinel crystals and it is in good agreement with literature [53]. The crystallite size for these characteristic peaks is found to be 46.5 nm. The XRD pattern for this catalyst is also found to conform with the COD on profile fitting of the peaks. The patterns for CoIO and CuIO prove that synthesized catalysts are highly crystalline and are in nano-scale size, thus confirming the success of solution combustion approach for synthesis of nano-catalysts. Both synthesized TMOs with above mentioned particle size (after grinding) were used in propellant formulations studied in this work.

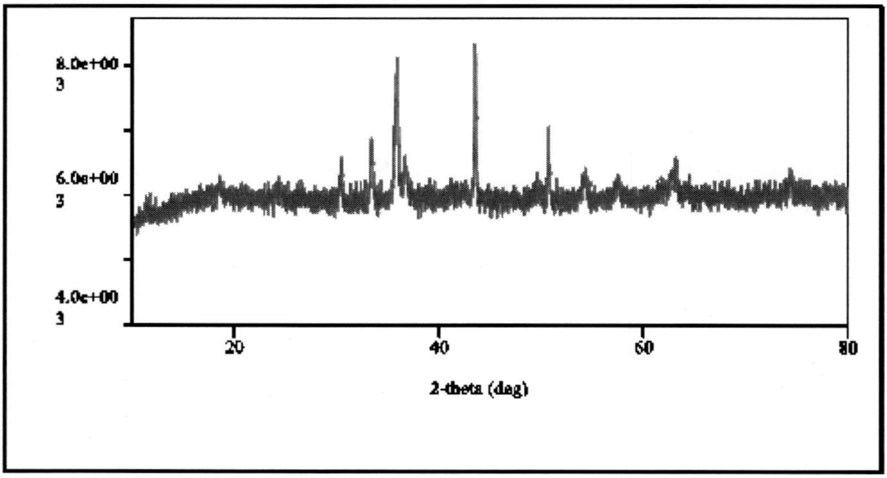

Figure 8. XRD pattern of CuIO

Scanning Electron Microscopic Study

SEM studies were carried out for both catalysts to understand their surface morphology, uniformity in shape and porous structure of sample obtained after solution combustion process without grinding. The SEM images were obtained at a resolution of 2000 x. The accelerating voltage used was 10 kV. The SEM images of CoIO and CuIO are given in Figures 9 and 10. The synthesized catalysts appear to have a similar surface morphology and show a lot of fine pores throughout the sintered network of solid crystalline materials. This feature makes it easier to grind it into finer particles and obtain a small grain size. The pore sizes appear to vary from 1-4 μm. SEM studies confirm the inference drawn from XRD analysis about the crystalline nature of catalysts.

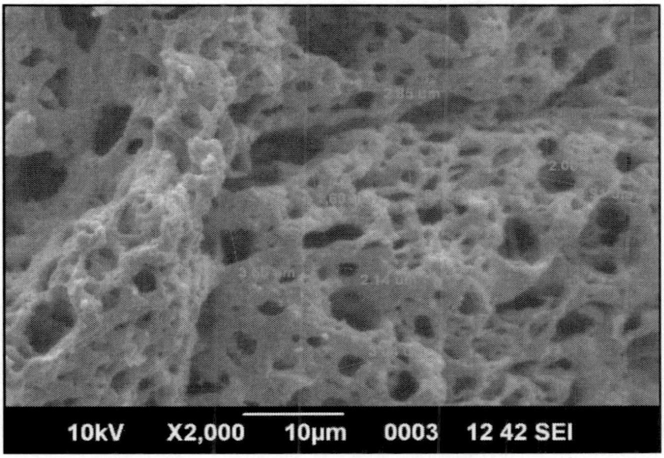

Figure 9. SEM image of CoIO.

Figure 10. SEM image of CuIO.

Fourier Transform Infrared Spectroscopy

Fourier transform infrared spectroscopy was performed to identify the nature of metal - oxygen bonds and atomic linkage in synthesized catalysts. FTIR was done for both samples in the range 4000 – 400 cm^{-1}.

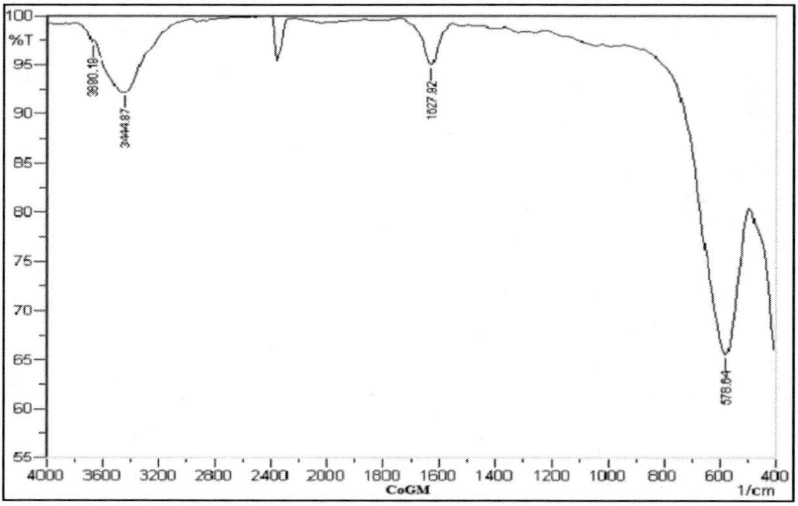

Figure 11. FTIR spectra of CoIO.

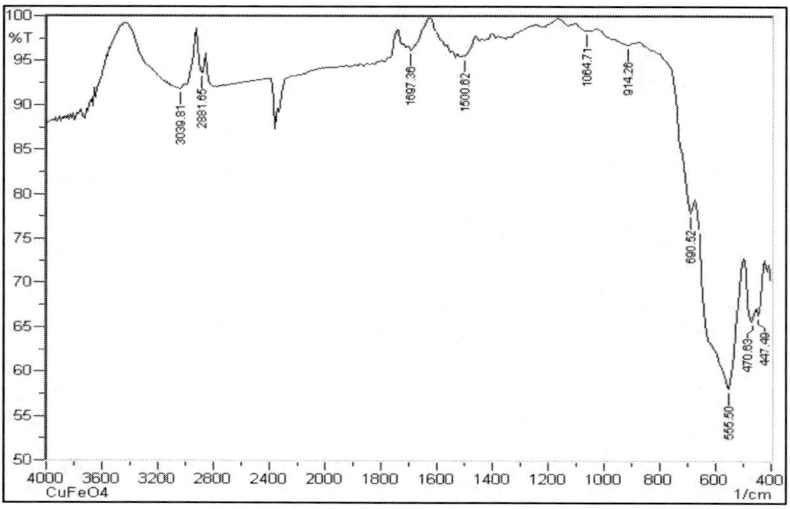

Figure 12. FTIR spectra of CuIO.

Cobalt ferrite ($CoFe_2O_4$) and Copper ferrite ($CuFe_2O_4$) are found to be the two catalysts synthesized in this study. It is reported that metal-ferrites [54] generally have a main metal oxygen band in the region 600 – 550 cm^{-1}. In both samples an absorption peak is observed in the given range as shown in Figures 11 and 12. This confirms that catalysts synthesized are Ferrites even though they may have different Co/Fe ratios (in case of CoIO). Thus, it can be concluded that solution combustion is a simple and efficient method to synthesize nano-bimetallic oxides.

THERMAL DECOMPOSITION STUDIES

Thermal decomposition studies were conducted to study the condensed phase behavior of composite solid propellants with and without catalysts. TMOs are known to catalyze the decomposition of AP and as a result enhance burn rate of composite solid propellants. Attempt has been made in this study to combine the positive catalytic effect of oxides of two transition metals (Co & Fe and Cu & Fe) in a single bimetallic TMO (CoIO and CuIO). In order to determine whether the catalysts synthesized in this study catalyze the decomposition of AP or the pyrolysis of the binder or both, decomposition studies have been carried out on catalyzed AP, HTPB binder and propellant samples. Similar experiments are conducted on pure AP, virgin HTPB and uncatalyzed propellant samples which are taken as reference against catalyzed samples for comparison purpose. The studies have also been carried out on commercially available TMO catalysts of corresponding individual metals (Fe_2O_3 and CuO) that are generally used in propellant formulations, so that the effect of bimetallic TMOs synthesized for this study can be compared. Digital Scanning Calorimetry (DSC), Thermogravimetric and Derivative-thermogravimetric Analysis (TGA/DTG) have been carried out in a Nitrogen environment in temperature range of 30°C – 700°C at a heating rate of 10°C/min using simultaneous thermal analyzer. The results obtained are discussed below.

Thermal Decomposition of AP

Pure AP decomposes in three stages as depicted in Figure 13. The first stage involves the transition of its crystalline structure from orthorhombic to cubic. This involves little to no mass loss and it is characterized by an endothermic peak at 246.3°C. The second stage is when there is a sharp drop in the slope of the TG curve (green) which indicates the onset of the major decomposition reaction. This is characterized by an exothermic peak at 294.6°C. The final stage is attained when about 98-100 wt.% of the decomposition is complete and it is characterized by a second exothermic peak at 414.3°C. The major mass loss (91 wt. %) occurs between 285°C (onset temperature) to 415°C. We define this temperature as onset temperature because the slope of the TGA curve drastically drops due to rapid loss of mass as can be seen from Figure 13. This onset temperature that characterizes the beginning of the major decomposition process occurs just before the first exothermic peak. The results are in good agreement with earlier works [25, 55].

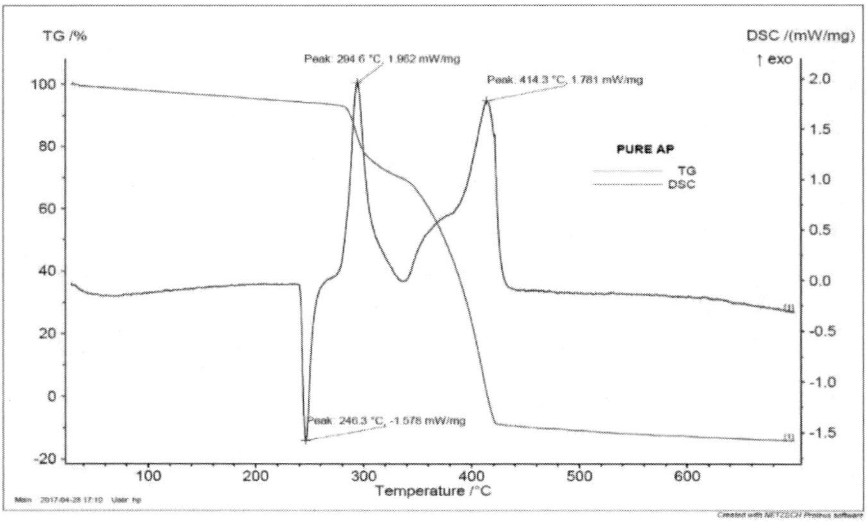

Figure 13. Thermal decomposition of pure AP.

A similar drop in slope can also be observed before the onset of the second exothermic peak, thus reconfirming the fact that there are two stages of AP decomposition, first being low temperature decomposition (LTD) and the other classified as high temperature decomposition (HTD). The thermogram is acquired to ascertain the influence of bimetallic TMOs on AP decomposition alone in conditions akin to condensed phase combustion.

Effect of Catalysts on the Thermal Decomposition of AP

Figure 14 shows the thermograms of pure AP and that of AP with 1%, 2% and 3% by wt. of CoIO. It can be seen from the plot that phase transition temperature of AP remains unaffected by the addition of CoIO. LTD peak also does not change noticeably upon addition of CoIO, however, there is a change in intensity of the peaks signifying increased heat flow during decomposition process.

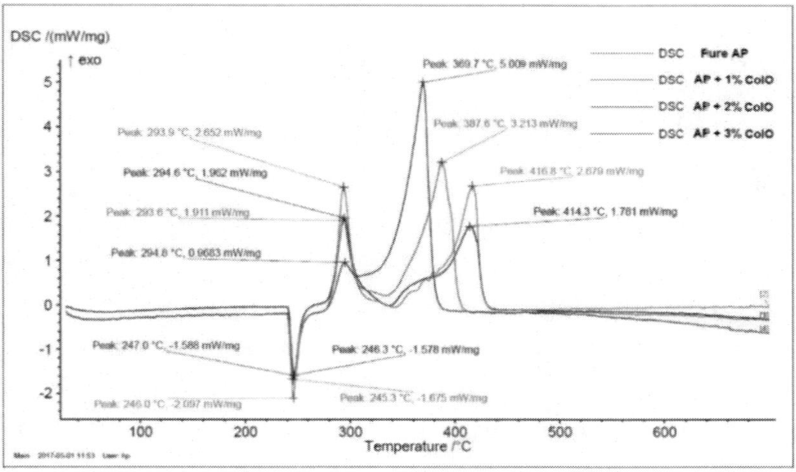

Figure 14. DSC Plot of pure AP and its catalyzed mixtures with CoIO at different concentration levels.

Addition of 1% CoIO results in increase in the height of all the peaks (endothermic, LTD & HTD). Addition of 2% CoIO does not result in the change in the endothermic peak height or LTD peak height but results in an increase in the peak height of HTD. Addition of 3% CoIO results in a decrease in the endothermic and LTD peak height and a significant increase in HTD peak height. The HTD temperature is also found to decrease with increase in the catalyst concentration implying that catalyst effectiveness improves with increase in its concentration.

The results show that addition of 3% CoIO is found to cause a 45°C decrease in the HTD temperature. This confirms that the HTD process is significantly influenced by the presence of higher concentration of catalyst. The increase in intensity of endothermic peaks may be attributed to an increased heat absorption which could be due to higher heat conductivity to the crystal lattice. The increase in the height of exothermic peaks signifies an increase in heat release which may be due to increased mass loss in that temperature range. Upon addition of 2% and 3% of CoIO, there is a significant change in the slope in DTG graph (not shown) which reveals an increase in the mass loss rate. Mass loss rates attain their peak values at 293.1°C & 296.3°C and 383.4°C & 364.0°C respectively. These values are found to be few degrees less than their corresponding LTD and HTD values recorded in case of pure AP.

The DSC plots for pure AP in comparison with AP + 1%, 2% and 3% by weight of CuIO catalyst are presented in Figure 15. The endothermic peaks do not show any measurable change upon addition of CuIO. The same trend is also observed in case of LTD temperature where the change is negligible. Further, HTD temperature is seen to decrease inversely with concentration of CuIO, meaning that highest catalytic activity is shown by the addition of 3% CuIO where the HTD temperature is found to decrease enormously i.e., by 79°C, however, in case of 2% and 3% addition of CuIO, the HTD peak appears to get split into two. This could be due to the change in the oxidation state of Cu1+ to Cu2+ after which the decomposition goes on to completion. The intensity of endothermic peak shows negligible difference on addition of CuIO which indicates that presence of CuIO does not influence thermal conductivity of the system. The change in intensity of the exothermic peaks can be attributed to improved reaction rate and resultant mass loss in corresponding regions of temperature.

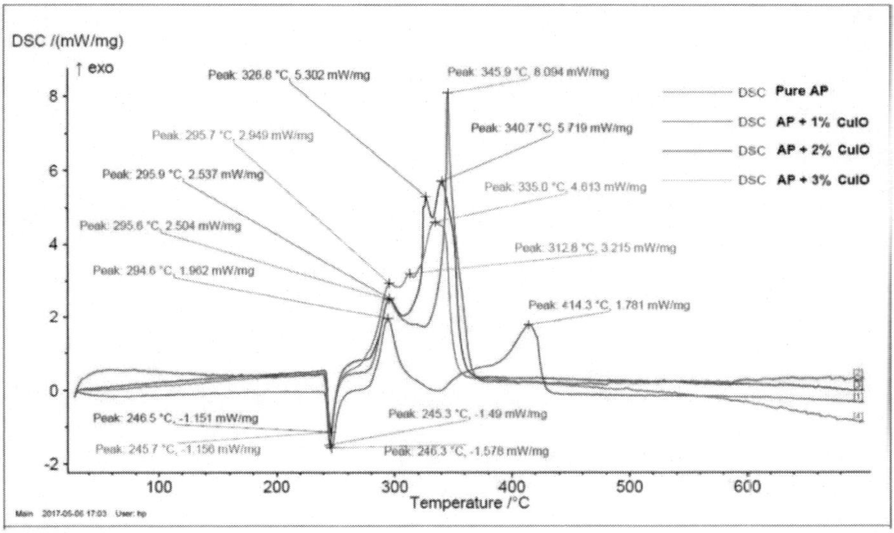

Figure 15. DSC Plot of pure AP and its catalyzed mixtures with CuIO at different concentration levels.

Further, the addition of CuIO does not cause measurable change in onset temperature of decomposition of AP. The temperature at which the decomposition goes on to completion decreases on the addition of CuIO and maximum reduction in end temperature occurs when 3 wt.% CuIO is added to AP. The rate of mass loss in corresponding DTG curves (not given) increases with increase in the concentration of CuIO, which also follows from the observation that there is a decrease in the end temperature. As observed for CoIO, the rate of mass loss reaches its peak value at a temperature that is a few degrees less than respective LTD and HTD values in this case as well. The only difference in this case is that there are two HTD peaks instead of one (in cases of 2% and 3% addition of CuIO). These peaks are separated by an average of about 15°C and the mass loss corresponding to these peaks is barely 1-2%.

Thermal Decomposition of Binder

The DSC results on thermal decomposition of cured binder with and without catalysts are shown in Figure 16.

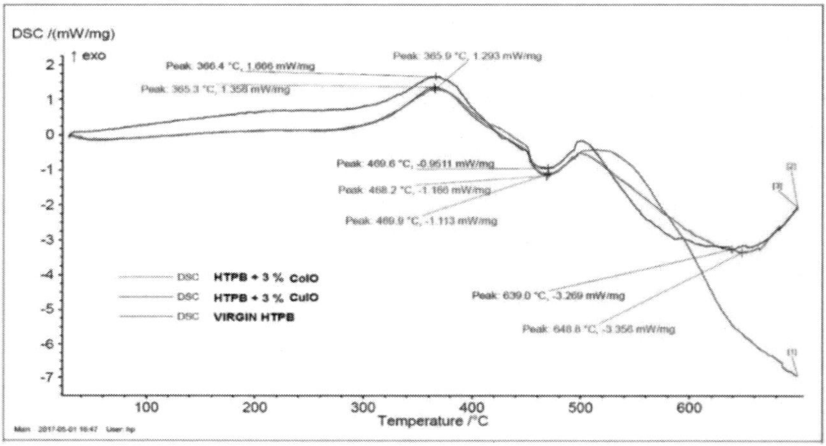

Figure 16. DSC Plot for virgin HTPB and HTPB embedded with CoIO & CuIO.

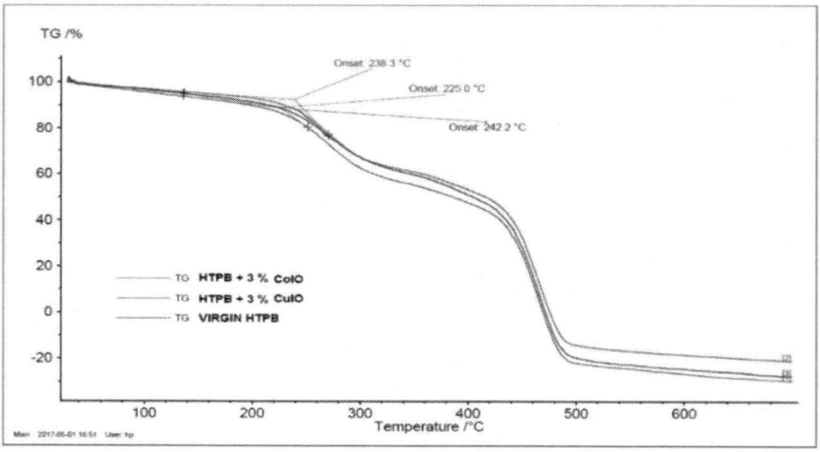

Figure 17. TG Plot for virgin HTPB and HTPB embedded with CoIO and CuIO.

From the curves, it may be inferred that addition of even 3 wt.% of catalysts (CoIO and CuIO) does not have any effect on the thermal decomposition of HTPB binder as far as the peak decomposition temperatures of first exothermic stage and subsequent endothermic stage is concerned, however, TG curves given in Figure 17 show the reduction in the onset temperature of HTPB decomposition from 242.2°C to 225°C in presence of CoIO catalyst. It may be inferred that catalysts influence the decomposition of TDI (that is used as curing agent for HTPB) possibly due to thermally reversible urethane reaction. TDI is known to evaporate in the range of 200-300°C [56].

Thermal Decomposition of CSPs

DSC and TGA studies were carried out on the virgin and catalyzed propellant samples to get an overall picture of the effect of synthesized catalysts. There is no observable change in the endothermic peak temperature upon addition of catalysts, however, the onset temperature of propellant is brought down by around 30°C for CoIO and CuIO catalyzed compositions at 1, 2 and 3 wt. % concentrations. The onset temperature for all catalyzed compositions is found to be in the range of 275 - 279°C and the concentration of the catalyst does not seem to alter this temperature (Figures 18 and 19).

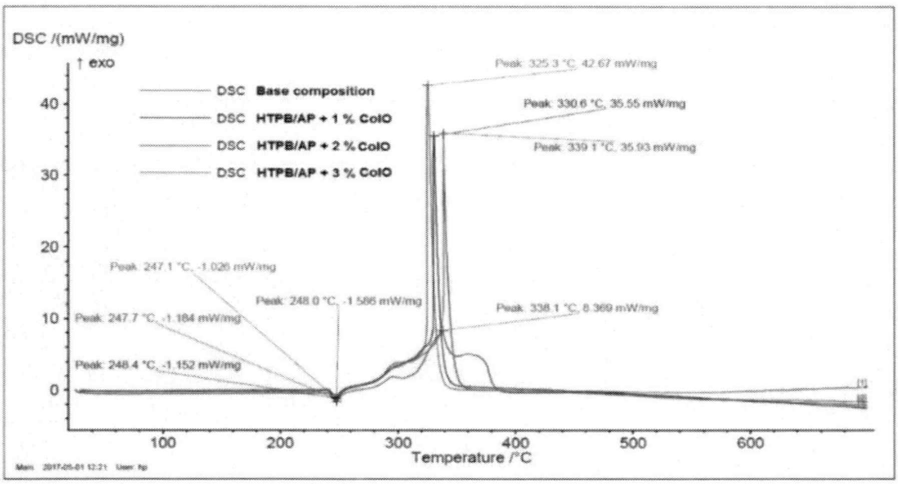

Figure 18. DSC Plot for base and CoIO catalyzed compositions at different concentration levels.

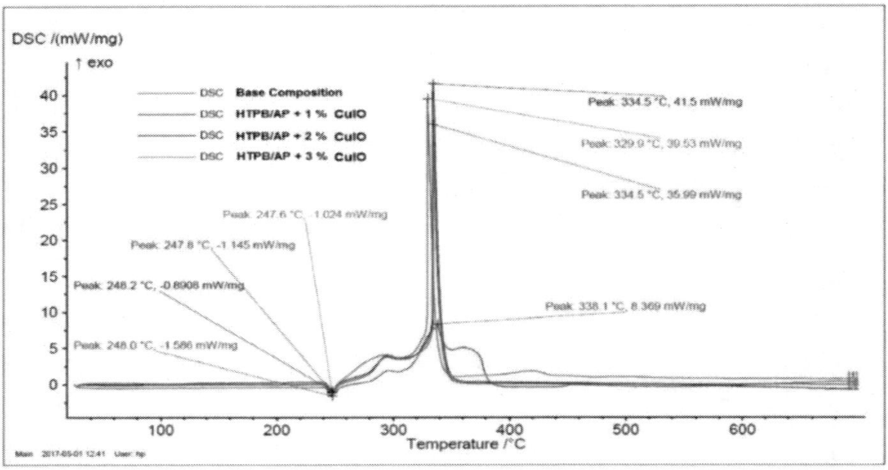

Figure 19. DSC Plot for base and CuIO catalyzed compositions at different concentration levels.

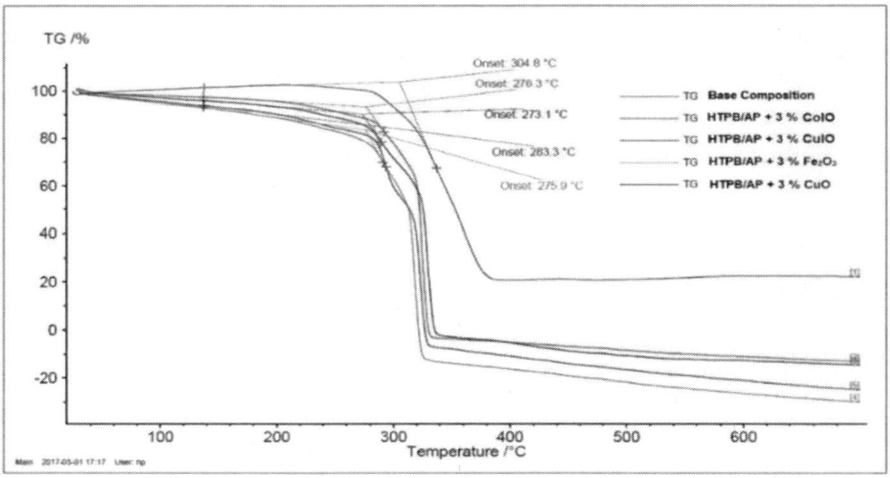

Figure 20. TG Plots for base propellant and others with conventional and synthesized catalysts.

The exothermic peak temperature is brought down by around 13°C for composition containing 3 wt. % of CoIO and by 10°C for the composition containing 3 wt. % of CuIO. Conventional catalysts, Fe_2O_3 and CuO lower the exothermic peak temperature by 20°C and 15°C respectively while they also do not cause any change in endothermic peak temperature. Their effect on lowering of onset temperature is same as that of the synthesized catalysts (Figure 20). The addition of aluminum powder (Al) and activated charcoal (AC) causes an increase in the exothermic peak temperature by 11°C (Figure 21) but onset temperature is, however, in the same temperature range as that of catalyzed non-aluminized propellant samples.

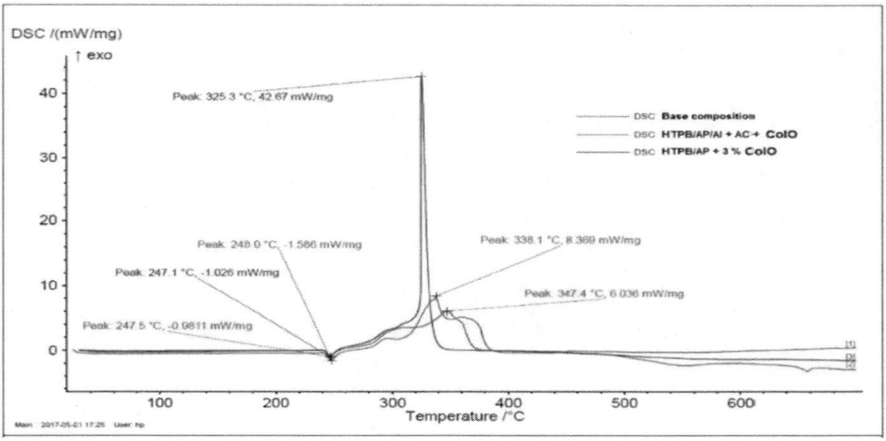

Figure 21. DSC Plot for base propellant and compositions containing CoIO, AC and Al.

BURN RATE STUDIES

This is the first study that explores the possibility of using nanosized bimetallic TMOs (CoIO and CuIO) in HTPB-AP based composite solid propellants for solid rocket propulsion. Burn rate studies have been conducted at three different pressures to study the variation of burn rate with pressure. Burn rates of virgin propellants are taken as reference against which the catalyzed propellants are compared. Three readings are taken for each propellant composition at a given pressure in order to establish repeatability (±1%). Burn rates of propellants catalyzed by CoIO and CuIO are recorded with combustion index in Table 3.

Table 3. Burn rates of virgin and catalyzed propellants containing CoIO and CuIO

Propellant Identity	Burn rate (mm/s) at			n	a
	2.07 MPa	4.14 MPa	6.21 MPa		
Base Composition	7.37	9.64	12.89	0.49	5.04
HTPB/AP + 1% CoIO	7.88	11.38	13.20	0.48	5.63
HTPB/AP + 2% CoIO	7.98	12.26	13.73	0.50	5.64
HTPB/AP + 3% CoIO	8.17	12.35	16.97	0.66	5.01
HTPB/AP + 1% CuIO	7.63	11.59	10.90	0.35	6.18
HTPB/AP + 2% CuIO	8.02	11.83	14.32	0.53	5.48
HTPB/AP + 3% CuIO	8.30	12.73	14.46	0.52	5.81

The results show that addition of CoIO and CuIO catalysts increases the burn rate of base propellant at all chamber pressures covered in this study. The burn rate is also found to increase with increase in the amount of CoIO added with the highest observed burn rate of 16.97 mm/s at 6.21 MPa for the propellant with 3 wt.% CoIO. The natural log of burn rate was plotted against the natural log of pressure to determine the values of pressure exponent or combustion index (n) of the propellant composition given in Table 3. The value of 'n' shows the pressure dependence of burn rate. The combustion index for the propellant sample with 1% CoIO (0.48) is found to be marginally less than that of virgin propellant (0.49) and the value of 'n' of the propellant sample with 2 wt.% CoIO is marginally higher (0.50). However, when the concentration of catalyst was increased to 3 wt.%, the 'n' value is found to increase significantly. This can be attributed to increased thermal conductivity of propellant surface due to decreased binder melt flow or phenomenal shift in rate controlling reactions during vapor phase combustion. A net increase of 31.65% in the burn rate of virgin propellant is observed when 3 wt.% of CoIO was added to it.

The addition of 2 wt.% and 3 wt.% CuIO is also found to increase the burn rate of the virgin propellant while only marginally increasing the value of 'n' (0.53 and 0.52 respectively). At 1 wt.% concentration, it is observed that burn rate increases up to a pressure of 4.14 MPa and then drops, thus exhibiting a dual burn rate index in the pressure range covered. The burn rate increments obtained by adding 2 wt.% and 3 wt.% of CuIO are 11.1% and 12.1% (at 6.21 MPa) respectively.

Burn rate studies for propellants catalyzed by 3% wt. of CuO and Fe_2O_3 were carried out to draw comparison between the effect of synthesized bimetallic TMOs of corresponding metals and the individual TMOs. Since Co_2O_3 cannot sustain combustion at high pressures [57], it was not considered for this comparative study. The experiments were carried out at 1.38 MPa, 2.07 MPa and 2.76 MPa to draw comparison between the effect of synthesized catalysts and the commercially available ones. The results in lower pressure segment are tabulated in Table 4.

From the results obtained, it can be concluded that order of efficacy of catalysts considered for this study are as follows, CoIO > Fe_2O_3 > CuIO > CuO. This establishes that CoIO catalyst synthesized in this study is superior to Fe_2O_3 which is established as one of the primary burn rate modifiers for composite solid propellants for years, at higher operating pressures. From the burn rate studies carried out for aluminized propellants containing activated charcoal and CoIO, it is found that the addition of Al powder and AC to catalyzed propellant composition can further enhance the burn rate (by 16%).

Table 4 Burn rates of catalyzed propellants containing conventional and synthesized catalysts

Propellant Identity	Burn Rate (mm/s) at			n	a
	1.38 MPa	2.07 MPa	2.76 MPa		
HTPB/AP	6.17	7.37	8.70	0.44	5.01
HTPB/AP + 3% Fe_2O_3	7.02	8.55	10.48	0.57	5.79
HTPB/AP + 3% CuO	6.93	8.39	9.66	0.48	5.94
HTPB/AP + 3% CuIO	6.80	8.30	10.03	0.56	5.64
HTPB/AP + 3% CoIO	7.46	8.17	11.38	0.58	5.93

CONCLUSION

Bimetallic nanosized TMOs can be successfully synthesized by solution combustion method and characterized using XRD, SEM and FTIR techniques. Particle size of the order of 45 ±2 nm can be obtained for both catalysts. Thermal decomposition studies conducted on AP and HTPB revealed that both CoIO and CuIO catalyze the decomposition of AP but have little effect on the pyrolysis of the binder in conditions related to condensed phase combustion. The results indicate that breaking of urethane linkage in presence of catalysts is influenced by advanced release of curing agent. This may affect the overall vapor phase combustion scenario, leading to higher burn rate as observed at higher combustion chamber pressures. It was found that addition of 3 wt. % of CoIO and CuIO, the decomposition temperature of AP was lowered by 44.6°C and 79.3°C respectively indicating that catalysts have a potential to significantly enhance condense phase combustion of CSP. Burn rate studies carried out at elevated pressures indicate that CoIO is more efficient in increasing the burn rate of the base propellant composition with a maximum observed burn rate of 16.97 mm/s observed at 6.21 MPa showing their pronounced efficacy in gas phase combustion of propellant as well. Burn rate studies performed on propellants catalyzed by commercially available catalysts indicate that the bimetallic TMOs synthesized in this study are superior burn rate modifiers to conventional TMOs. CoIO is found to be a superior burn rate catalyst than commercial Fe_2O_3 which has been regarded as the best burn rate catalyst for years. The pressure exponent is found to increase with increase in concentration of CoIO to a maximum observed value of 0.66 at 3 wt.% level, however, CuIO has shown little influence on combustion index when compared to base composition. Addition of aluminum powder and activated charcoal can further enhance the burn rate of the catalyzed propellants containing CoIO by 16%.

ACKNOWLEDGMENTS

The support provided by Dr. Narendra Yadav, Research Scientist in the Department and the staff of the Central Instrument Facility of the Institute is gratefully acknowledged. The authors also wish to thank Bernard M. Kosowski, President and Dr. Tony Manzara, VP Technology, Mach I. Inc, PA, USA for their help and cooperation in editing the text.

REFERENCES

[1] Sutton G.P., *Rocket Propulsion Elements,* Wiley, New York, (1992). p. 426 ISBN-0-471-32642-9.

[2] Gupta, G., Jawale, L., Mehilal, Bhattacharya, B. Various Methods for the Determination of the Burning Rates of Solid Propellants - An Overview, *Central European Journal of Energetic Materials,* (2015). 12(3), 593-620.

[3] Crawford B.L., Hugget C., Daniels F., Wilfong R.E., Direct Determination of Burning Rates of Propellant Powders, *Anal. Chem.,* (1947). 19(19), 630-633.

[4] Grune D., Kegler W., Nicolas M., Benhaim P., Studies of the Combustion of Solid Propellants at High Pressures, (1977). *Proc. Int. Symp. Ballist.,* 3rd, A2: 12.

[5] Akira I., Kiroku Y., Kazuo K., Shoichiro A., Isao K., *Burning Rate Measurement of Solid Propellants by Means of Phototransistors* (in Japanese), Kogyo Kagaku Zasshi, (1962). 65, 1218-1219.

[6] Lui F., A New Method for Measurement of Burning Rate of Propellant and Explosives, in: *Combustion and Detonation Phenomena, 19th Int. Annu. Conf. ICT,* Karlsruhe, (1988). 49/1-49/11.

[7] Wood H.L., O' Brien W.F., Dale C.B., (1983). Measurement of Solid Propellant Burning Rates Employing Microwave Techniques, *20th JANNAF Combustion Meeting.*

[8] Song S.J., Kim H.J., Ko S.F., Oh H.T., Kim I.C., Yoo J.C., Jung J.Y., (2009). Measurement Various Methods for the Determination of the Burning Rates of Solid Propellants... 617 of Solid Propellant Burning Rates by Analysis of Ultrasonic Full Waveforms, *J. Mech. Sci. Technol.,* 23(4), 1112-1117.

[9] Jeffery J.M., Herman K., (2002). Evaluation of Ultrasound Technique for Solid-propellant Burning-rate Response Measurements, *J. Propul. Power,* 18(3), 641-651.

[10] Bozic V.S., Blagojevic D.D., Anicin B.A., (1997). Measurement System for Determining Solid Rocket Propellant Burning Rate using Transmission Microwave Interferometry, *J. Propul. Power, 13*(4), 457-462.

[11] Masahiro K., Angelo V., Shin-Ichiro T., (1997). X-ray Diagnostics for Local Burning Rate Measurement of Solid Propellants, *Proc. Int. Symp. Spec. Top. Chem. Propulsion,* Stockholm, 918-926.

[12] Osborn J.R., Bethel H.E., (1964). Technique for Measuring Burning Rates of Solid Propellants, *Rev. Sci. Instrum.,* 35(9), 1130-1134.

[13] Eisenreich N., Kugler H.P., Sinn F., (1987). An Optical System for Measuring the Burning Rate of Solid Propellant Strands, *Propellants Explos. Pyrotech.,* 12(3), 78-80.

[14] Kubota N., (2002). Propellants and Explosives: Thermochemical Aspects of Combustion. *Measurements of Burning Rate and Combustion Wave Structur*e, Wiley-VCH, 235-236, ISBN: 9783527302109.

[15] Wang J., Gao Z., Sang B., (2001). A Laser Technology for Measurement of Solid Propellant Transient Burning Rates During Rapid Depressurization, *Fuel,* 80(2), 263-271.

[16] Caveny L.H., Pokrocos L.M., Lelshain G.R., (1977). *High Pressure Burning Rates of Multibase Propellants,* Department of Aerospace and Mechanical Sciences, Princeton University, New Jersey, Report AMS-1377, 75.

[17] Brooks W.T., (1970). A Method for More Reproducible Burning Rate Determination, *J. Spacecraft Rockets*, 7(12), 1488-1489.

[18] Whitney C.K., Owens T.F., Paskind J., Rubin M.B., (1965). *Scout Motor Performances and Prediction Study (PAPS),* NASA-CR-336, NASA Langley Research Center, USA.

[19] Summerfield M., Sutherland G.S., Webb M.J., Taback H.J., Hall K.P., (1958). *Burning of Ammonium Perchlorate Propellants,* American Rocket Society, New York, 737-758.

[20] DeLuca, L.T., Shimada T., Sinditskii, V.P. and Calabro M. (Eds.), (2017). *Chemical Rocket Propulsion A Comprehensive Survey of Energetic Materials*, Springer, Switzerland, 3-18 ISBN 978-3-319-27746-2.

[21] Satyanarayanan R. Chakravarthy, Edward W. Price, and Robert K. Sigman. (1997). Mechanism of Burning Rate Enhancement of Composite Solid Propellants by Ferric Oxide, *Journal of Propulsion and Power,* 13: 4, 471-480.

[22] Ishitha, K., and Ramakrishna, P.A., (2014). Studies on the role of iron oxide and copper chromite in solid propellant combustion. *Combust. Flame*, 10, 2717–2728.

[23] Krishnan, S., and Swami, R.D., (1998). Effect of burning rate modifiers on sub-atmospheric flame temperature of AP/HTPB composite solid propellant. *Def. Sci. J.,* 48:2, 211.

[24] Patil, P.R., Krishnamurthy, V.N., and Joshi, S.S., (2006). Differential scanning calorimetric study of HTPB based composite propellants in presence of nano ferric oxide. *Propellants Explos. Pyrotech.*, 31:6, 442.

[25] Padwal, M.B., Varma, M. (2018). Thermal decomposition and combustion characteristics of HTPB-coarse AP composite solid propellants catalyzed with Fe_2O_3, *Combustion Science and Technology,* 190:9, 1614-1629.

[26] Campos, E.A., Dutra, R.C.L., Rezende, L.C., Diniz, M.F., Nawa, W.M.D. and Iha, K., (2010). Performance evaluation of commercial copper chromites as burn rate catalyst for solid propellant, *J. Aerosp. Technol.,* 2:3, 323-330.

[27] Rajeev, R., Devi, K.A., Abraham, A., Krishnan, K., Krishnan, T.E., Ninan, K.N. and Nair, C.G.R., (1995). Thermal decomposition

studies. Part 19. Kinetics and mechanism of thermal decomposition of copper ammonium chromate precursor to copper chromite catalyst and correlation of surface parameters of the catalyst with propellant burning rate, *Thermochimica Acta,* 254, 235-247.

[28] Gore, G.M., Tipare, K.R., Bhatewara, R.G., Prasad, U.S., Gupta, M. and Mane, S.R., (1999). Evaluation of Ferrocene Derivatives as Burn Rate Modifiers in AP/HTPB-Based Composite Propellants, *Def. Sc. J.*, 49:2, 151-158.

[29] Sinditskii, V.P., Chernyi, A.N., Marchenkov, D.A., (2014). Mechanisms of combustion catalysis by ferrocene derivatives. 1. Combustion of ammonium perchlorate and ferrocene, *Combust. Explos. And Shock Waves,* 50:1, 51-59.

[30] Krishnan, S. and Jeenu, R., (1992). Combustion characteristics of AP/HTPB propellants with burning rate modifiers, *Journal of Propulsion and Power,* 8:4, 748-755.

[31] Ghorpade, V.G., Dey, A., Jawale, L.S., Kotbagi, A.M., Kumar, A., Gupta, M. (2010). Study of Burn Rate Suppressants in AP-Based Composite Propellants, *Prop. Explos. And Pyrotech.,* 35:1, 53-56.

[32] Dubey R., Chawla M., Siril P.F., Singh G. (2013). Bi-metallic nanocomposites of Mn with very high catalytic activity for burning rate enhancement of composite solid propellants, *Thermochimica Acta,* 572:30.

[33] Li W, Cheng H. (2007). Cu–Cr–O nanocomposites: synthesis and characterization as catalysts for solid state propellants. *Solid State Sci,* 9:750.

[34] Yetter R.A., Risha, G.A., Son S.F., (2009). Metal particle combustion and nanotechnology. *Proc. Combus. Inst.,* 32: 1819–1838.

[35] Hosseini S.G., Abazari R, Gavi A. (2014). Pure CuCr2O4 nanoparticles: synthesis, characterization and their morphological and size effects on the catalytic thermal decomposition of ammonium perchlorate. *Solid State Sciences,* 37:72-79.

[36] Gore G.M., Nazare A.N., Divekar C.N., Hait S.K., Asthana S.N., (2004). Studies on nano-aluminized high burning rate AP-composite propellants, *J. Energ. Mater.,* 22: 151.

[37] Kishore, K., (1979). Comprehensive view of the combustion models of composite solid propellants, *AIAA J.,* 17:11, 1216-1224.

[38] Guirao, C., Williams, F.A., (1971). A model for ammonium perchlorate deflagration between 20 and 100 atm, *AIAA J.,* 9:1345-1356.

[39] Johnson, W.E., Nachbar, W. (1962). Deflagration limits in the steady linear burning of a monopropellant with application to ammonium perchlorate, *VIIIth Symposium (International) on Combustion,* The Combustion Institute, Pittsburgh, USA, 679-689.

[40] Kishore K., Sridhara, K., (1999). *Solid Propellant Chemistry: Condensed phase behavior of ammonium perchlorate based solid propellants,* DRDO Monograph, DESIDOC, 31-70.

[41] Beck, W.H., (1987). Pyrolysis studies of polymeric materials used as binders in composite propellants: a review, *Combust. and Flame,* 70:171-190.

[42] Summerfield, M., Sutherland, G.S., Webb, M.J., Taback, H.J., Hall, K.P., (1960). *Burning mechanism of ammonium perchlorate propellants, Solid Propellant Rocket Research,* Academic Press, New York, 141-182.

[43] Hermance, C.E., (1966). A model of composite propellant combustion including surface heterogeneity and heat generation, *AIAA J.,* 4: 1629-1637.

[44] Beckstead, M.W., Derr, R.L., Price, C.F., (1970). A model of composite solid propellant combustion based on multiple flames, *AIAA J.,* 8: 2200-2207.

[45] Glick, R.L., (1976). Distribution functions for statistical analysis of monodisperse composite solid propellant combustion, *AIAA J.,* 14: 1631-1633.

[46] Knott, G.M., Brewster, M.Q., Jackson, T.L., (1998). Simplified modeling of composite propellant combustion with finite Peclet

number," *35th JANNAF Combustion Subcommittee Meeting,* CPIA Pub. No. 680, 1: 655.

[47] Wang, X., Jackson, T.L., Massa, L., (2004). Numerical simulation of heterogeneous propellant combustion by a level set method, *J. Combust. Theory Model.,* 8:227.

[48] Cai, W., Yang, V.A., (2000). *Model of AP/HTPB composite propellant combustion,* AIAA Paper No. 2000-0311.

[49] Jeppson, M.B., Beckstead, M.W., Jing, Q., (1998). *A kinetic model for the premixed combustion of a fine AP/HTPB composite propellant,* AIAA Paper No. 98-0447.

[50] Beckstead, M.W., Puduppakkama K., Thakre, P., Yang, V., (2007). Modeling of combustion and ignition of solid-propellant ingredients, *Progress in Energy and Combustion Science,* 33: 497.

[51] Kishore, K., (1979). Polymer flame inhibition in presence of transition metal oxides, *Colloid and Polym. Sci.*, 257:984.

[52] Boldyrev, V.V., (2006). Thermal decomposition of ammonium perchlorate, *Thermochimica Acta,* 443: 1.

[53] Hosseini, S.N., Enayati, M.H., Karimzadeh, F., Sammes, N.H., (2015), Synthesizing $CuFe_2O_4$ spinel powders by a combustion like process for solid oxide fuel cell interconnect coatings, *Int. J. Chem. Mol. Nuclear Mat. and Metallurgical Eng.* 9:7, 857-860.

[54] Kasapoglu, N., Baykal, A., Koseoglu, Y., Toprak, M.S. (2007), Microwave-assisted combustion synthesis of $CoFe_2O_4$with urea and its magnetic characterization, *Scripta Materialia* 57, 441-444.

[55] Ma, Z., Li, F., Bai, H. (2006), Effect of Fe2O3 in Fe2O3/AP composite particles on thermal decomposition of AP and on burning rate of the composite propellant, *Propellants Explos. Pyrotech.,* 31, 447.

[56] Ranjbar, Z., Montazeri, S., Raei Nayini, M.M., Jannesari, A. (2010), Synthesis and characterization of diethylene glycol monobutyl ether-blocked diisocyanate crosslinkers, *Prog. Org. Coat.* 69, 426.

[57] Kishore, K., Pai Vernekar, V.R., Sunitha, M.R. (1980), Action of transition metal oxides on composite solid propellants, *AIAA J.*, 18:11, 1404-1407.

In: Thermal Decomposition
Editor: Kalyna Jocosa Pace
ISBN: 978-1-53615-460-3

Chapter 3

THERMAL RUNAWAY IN ALKALINE BATTERIES: PROCESS AND EFFECTS

N. E. Galushkin*, N. N. Yazvinskaya and D. N. Galushkin
Department of Technics and Techology,
Don State Technical University,
Laboratory of Electrochemical and Hydrogen Energy,
Shakhty, Russia

ABSTRACT

This chapter discusses the thermal runaway in alkaline batteries. It was shown that a lot of experimental data exists that contradict the generally accepted thermal runaway mechanism. On the basis of all the available experimental data, there was proposed the new thermal runaway mechanism. There it was shown that there are two processes of accumulation that step by step bring the nickel-cadmium batteries to the thermal runaway. First, this is a process of hydrogen accumulation inside

* Corresponding Author's E-mail: galushkinne@mail.ru.

of battery electrodes during their operation. Second, this is a process of dendrites accumulation on the cadmium electrodes of the batteries. It was experimentally proved (using thermal decomposition of sintered electrodes) that during Ni-Cd batteries long service life (more than five years), the hydrogen accumulation in large quantities takes place in a form of nickel hydrides in a sintered nickel matrix of oxide-nickel electrodes. In this chapter, it was deduced from experiments that an exothermic reaction in the thermal runaway process in alkaline batteries is the electrochemical reaction of atomic hydrogen recombination $H_{ads\ \mathrm{Cd}} + H_{ads\ \mathrm{Ni}} \rightarrow H_2 \uparrow$ ($H_2O + H_{ads} + e^- \rightarrow H_2 \uparrow + OH^-$ on a cathode and $H_{ads} + OH^- \rightarrow H_2O + e^-$ on an anode). A rate-limiting step for this reaction on both a cathode and an anode is found to be a step of metal-hydrides disintegration. This electrochemical reaction proceeds at voltage 0.5–0.6 V on battery terminals. An analytical model of thermal runaway in alkaline batteries is proposed.

Keywords: thermal decomposition, thermal runaway, metal hydride, alkaline batteries, hydrogen accumulation

1. Introduction

Thermal runaway occurs in nickel-cadmium, lead–acid, lithium-ion, nickel-metal hydride, etc., batteries [1], which is to say that the thermal runaway is a phenomenon inherent in batteries of practically all electrochemical systems.

In outward appearance, the thermal runaway, in the batteries of all the above systems, runs its course in the same way. In the case of batteries recharging under constant voltage or in a case of their work in floating service, they can suddenly be heated up to a great extent, fuse, be on fire and steam or even burst depending on their electrochemical system, structure, housing material, etc.

At present time, the batteries in which the thermal runaway is observed are installed in many devices of both general and special-purpose: wireless phones, computers, airplanes, back-up sources of communication networks, etc. Inevitably the battery thermal runaway in those devices and systems is going to result in either their breakdown or in difficulties in their work. Hence, nowadays the thermal runaway is a serious obstacle in the proper functioning of lots of modern devices and systems.

In spite of the importance of the above problem, studies are desperately scant in world literature dedicated to investigation of this dangerous and interesting phenomenon. At the present time, this phenomenon is studied quite intensively in lithium-ion batteries [2-8], but desperately seldom in batteries with aqueous electrolyte [9, 10]: especially few such studies are there for alkaline batteries [1, 11]. Meanwhile the alkaline batteries are an integral part of electric systems of airplanes, electric vehicles, railway transportation, etc.

Such disregard to this blusterous phenomenon can be explained, in our opinion, by only two reasons. First, the thermal runaway is a very rare phenomenon and hence does not jeopardize every-day work of devices and systems. That is why battery producers do not invest considerable finance into this problem's investigation. Second, the thermal runaway mechanism seems to many engineers just evident [1], while until now there have been neither direct experimental evidences of this mechanism nor its artificial reproduction. There also has not yet been conducted, a detailed analysis of thermal runaway products (except of the lithium batteries).

According to the standard point of view [1], the thermal runaway in batteries goes on with the following scenario. In a case of a battery long recharging under constant voltage (or at its floating service), heating-up takes place, which results in the decreasing of a battery's internal resistance and an increase of recharging current, which in turn magnifies heating up, etc. So the thermal runaway is a result of a positive feedback between a current and a temperature of batteries during their charging under a constant voltage. Besides, for hermetic batteries, there is a belief that an essential initial contribution to batteries heating-up is made by exothermic reactions of an oxygen cycle.

In a case of thermal runaway initiation, a battery heats up until high temperature values; an electrolyte evaporates instantly; a battery plastic body melts and sometimes it catches fire or explodes depending on the battery body construction and material.

So the thermal runaway classical mechanism supposes that the reason for the thermal runaway is an acceleration of a battery charge reaction (on account of its heating-up)

$$Ni(OH)_2 + OH^- \rightarrow NiOOH + H_2O + e^- \quad (1)$$

$$Cd(OH)_2 + 2e^- \rightarrow Cd + 2OH^- \quad (2)$$

or an acceleration of an electrolyte decomposition reactions

$$2e^- + 2H_2O \rightarrow H_2 + 2OH^- \quad \text{(cathode)} \quad (3)$$

$$2OH^- \rightarrow 1/2O_2 + H_2O + 2e^- \quad \text{(anode)} \quad (4)$$

as the thermal runaway takes place in cases of long-time battery recharging.

In our previous papers [12-18] it was shown that a lot of experimental data exists that contradict the generally accepted thermal runaway mechanism.

First, according to the classical thermal runaway mechanism [1], a probability of the thermal runaway initiation should not depend on a battery's operating life. Nevertheless, the available direct experimental data show that the thermal runaway initiation probability depends on the battery's operating life to a large extent [15]. Please note that in new Ni-Cd batteries, the thermal runaway is not encountered at all.

Second, according to the classical mechanism [1], the thermal runaway is connected with a battery charging reaction (1,2) intensification and an electrolyte decomposition (3,4) (in the case of a battery recharging) because of the battery heating up. Therefore, at the thermal runaway from the battery, steam and a small amount of hydrogen and oxygen in proportion 2:1 can be released. But the experimental measures [12] show that as a result of a thermal runaway from a battery, steam (in amount approximately equal to an amount of an electrolyte in the battery) and gas (95 to 97% of which is represented by the hydrogen) are released. For example, at the thermal runaway from the battery KSX-25 (with sintered electrodes by capacity 25 Ah), approximately 270 liters hydrogen and 14 liters oxygen are released [12]. Notably, the released hydrogen amount exceeds its amount in the whole electrolyte if the latter would be decompounded on hydrogen and oxygen.

Third, according to the classical mechanism [1], the thermal runaway occurs because of an energy obtained by a battery from a recharger. But the direct experimental studies show that as a result of a thermal runaway, a battery releases approximately 140 times more energy than it obtains from the recharger [17]. Therefore, the thermal runaway is connected with the development of a powerful exothermic reaction. Though the thermal runaway classical mechanism does not suppose a development of some additional electrochemical reactions, except for one of battery charging and one of electrolyte decomposition (at battery recharging). Besides, there are a lot of other experimental facts contradicting the thermal runaway classical mechanism [12-18].

2. Thermal Runaway in Alkaline Batteries

2.1. Influence of External Factors on the Probability of Thermal Runaway

First of all, let's investigate the impact of some factors on the probability of thermal runaway development. Among those chosen factors there is a charge voltage, an environmental temperature and a battery operating life. For the study, the batteries KSX-25 were used. This battery is installed in onboard systems of aircraft of many types as a part of the battery 20KSX-25.

For the batteries' cycling process, a recharger was used which allowed for installing one of a number of fixed charge voltages: 1.45; 1.67; 1.87; 2.2 V. The recharger allowed working on a constant basis with currents up to 300 A and on short-duration basis with currents up to 1000 A.

In order to obtain more voluminous statistics for less of a period of time, the recharger was connected to a group of ten parallel batteries in a hard metal clamp band. These batteries joining up in parallel was made with a help of two heavy duty metal runs, to which (separately) batteries positive and negative terminals were screwed on.

In order that a thermal runaway developed in one battery had no influence on the probability of a thermal runaway development in neighboring batteries (which could happen because of their additional heat-up), heat-insulating wood block fillers, two centimeters thick were inserted between the batteries in the metal clamp-band.

In the first group of experiments, the impact of the recharger voltage on the thermal runaway development probability was investigated. In connection with the experiments in this study, 40 batteries were taken from two batteries 20KSX-25. The chosen batteries were of the type KSX-25 and their life in operation was seven years. In a random manner, those batteries were divided into four groups with ten items in each. The first group of the batteries was charged under constant voltage 1.45 V; the second one 1.67 V; the third one 1.87 V and the last one 2.2 V. The lowest value of the studied charge voltage range corresponds to an average working voltage in floating service, when the battery is a part of the battery 20KSX-25 onsite. Batteries of the groups were charged and discharged eighty times each. Thus, 10*80=800 charge-discharge cycles were made at each value of the charging voltages.

The charging of each group was conducted under constant voltage corresponding to this group during ten hours. The discharging was implemented according to the operation manual for the battery 20KSX-25. It was made by the current 10 A up to voltage value 1 V on the terminals of the batteries. The cycling of the batteries was fulfilled under environmental temperature 25°C. The study results are shown in Table 1.

The results of the batteries KSX-25 cycling (Table 1) obviously show that the probability of the thermal runaway rises with an increase of the battery charge voltage.

In the second group of experiments, the environmental temperature impact on the thermal runaway development probability was studied. For this study, the batteries were cycled in a thermal chamber at the temperature 40°C and 65°C. It is necessary to note that different groups of batteries were used for each group of experiments and at each value of the temperatures. The results of the batteries cycling are shown in Tables 2 and 3.

The results of batteries KSX-25 cycling (Tables 2 and 3) obviously show that the probability of the thermal runaway rises with the environmental temperature increase. These experiments justify recommendations on batteries KSX-25 operation that prohibit work of the batteries in floating service under voltage more than 1.5 V per cell and at the temperature more than 45°C.

Table 1. Results of batteries KSX-25 cycling at environmental temperature 25°C

Batteries group number	**1**	**2**	**3**	**4**
Charge voltage (V)	1.45	1.67	1.87	2.2
Period of operation (years)	7	7	7	7
Number of charge-discharge cycles	800	800	800	800
Number of thermal runaways	0	0	1	3

Table 2. Results of batteries KSX-25 cycling at environmental temperature 40°C

Batteries group number	**1**	**2**	**3**	**4**
Charge voltage (V)	1.45	1.67	1.87	2.2
Period of operation (years)	7	7	7	7
Number of charge-discharge cycles	800	800	800	800
Number of thermal runaways	0	0	2	5

Table 3. Results of batteries KSX-25 cycling at environmental temperature 65°C

Batteries group number	**1**	**2**	**3**	**4**
Charge voltage (V)	1.45	1.67	1.87	2.2
Period of operation (years)	7	7	7	7
Number of charge-discharge cycles	800	800	800	800
Number of thermal runaways	0	1	5	13

In the third group of experiments, the impact of the batteries operating life on the thermal runaway development probability was investigated. In these experiments we used: new batteries, batteries with an operating life of 2, 5 and 7 years. All batteries were operated under the same conditions. The cycling results are shown in the Table 4.

Table 4. Results of batteries KSX-25 cycling with different operating life at temperature 40°C and constant charge voltage 2.2 V

Batteries group number	**1**	**2**	**3**	**4**
Period of operation (years)	New	2	5	7
Number of charge-discharge cycles	800	800	800	800
Number of thermal runaways	0	0	1	5

The results of the cycling (Table 4) show that the probability of the thermal runaway grows with the increase of the batteries operating life. Along with it, in the case of the new batteries, with a small operating life, the thermal runaway is either impossible or extremely unlikely.

This experiment shows for sure that among factors responsible for the thermal runaway, there are some ageing processes which increase the probability of the thermal runaway with the batteries' operation in progress.

2.2. Analysis of the Change in Parameters in Ni-Cd Batteries during Thermal Runaway

In the experiments, vented Ni-Cd batteries KSX-25 with sintered plates by capacity 25 Ah were used. In our previous Section 2.1, it was shown, that the probability for thermal runaway increases with the service life of the battery and an increase in charging voltage. Hence, batteries with a service life of over 6 years were used for this study. Charging was performed under a voltage of 1.87 V for 10 hours. Discharge was performed according to the operation manual of the KSX-25 battery under a current of 10 A to a voltage of 1 V. Charging and discharging was performed for a pack of ten batteries connected in parallel. This increased the number of experiments during the same time period and therefore the possibility of thermal runaway also increased.

Wooden thermal insulating pads, two centimeters thick and clamped in a metal brace were inserted between the batteries to exclude the possibility of thermal runaway initiating in one battery to influence or initiate the thermal runaway in neighboring batteries due to additional overheating.

The voltage was measured separately on the terminals of each battery. Charge current was measured with a shunts resistance of 0.001 Ohm and maximum current of 500 A. Shunts were included in a circuit separately for each battery between a negative terminal and a common bus. The temperature was measured with aid of thermocouples connected to positive terminals of each battery. Via analog-to-digital converter, all data were delivered to the computer. The steam and gas which evolved as a result of the thermal runaway got out freely into a flexible reservoir 1060 L in volume. Figure 1 shows the parameters for two batteries during the process of thermal runaway.

Let's consider some visual consequences of the thermal runaway. On Figure 2 the electrodes are shown of two batteries exposed to the process of thermal runaway. From the visual analysis of Figure 2, the following conclusions can be made.

First, because of the thermal runaway, a separator is burnt through only in certain places but not on the whole electrode surface. The conclusion can therefore be made that the thermal runaway is a local phenomenon.

Second, as the circles of burning-through of the separator on the electrodes are located in various places at different batteries exposed to the thermal runaway, the conclusion can be made that the process of initiation of the thermal runaway lacks any regular pattern and occurs at random and in absolutely incidental places.

Third, due to the fact that the places of burning-through of the separator possess a shape of regular circles and also an absolutely different radius, one can suppose that the thermal runaway initiates from a point of a surface and then (like a combustion process) expands radially in a uniform manner from this point.

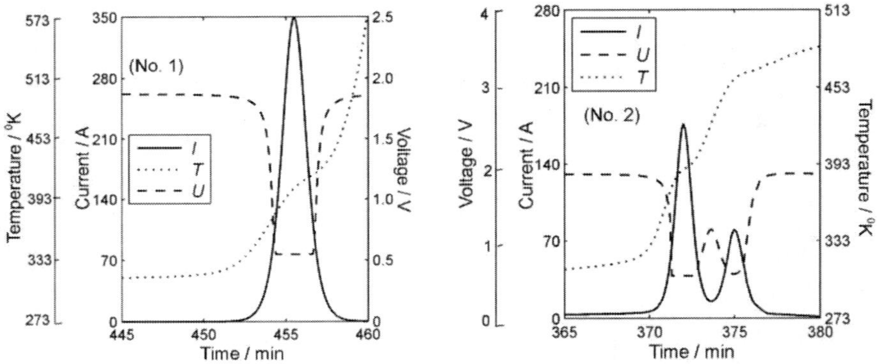

Figure 1. Change in parameters of the KSX-25 batteries (No. 1,2) during thermal runaway: *I* is the charging current of the battery, *U* is the voltage of the battery terminals, *T* is the temperature of the battery positive terminal.

Fourth, as the circles of burning-through of the separator do not spread on the entire electrode surface and since all of them have a different random radius, a conclusion can be made that in the course of the thermal runaway, some processes exist which prevent its propagation on the whole electrode and assist in an extinction of the thermal runaway.

Finally, in the experiment there were two current bumps (Figure 1 (No. 2)) and exactly two areas of burning-through of the separator on the electrodes corresponding to them (Figure 2 (No. 2)). Hence, during one charging, a thermal runaway can be initiated and die out repeatedly. That is why the following mechanism can be proposed for initiation and extinction of the thermal runaway. The initiation and the extinction of the thermal runaway in any place of an electrode causes first a fast growth of a charge current and then an equally fast current downfall because of an electrolytic solution evaporation in this place and of vapor locking in the burnt hole of the separator. In its turn this would result in enlargement of an average current density in other areas of the electrodes which can launch the thermal runaway in another place, probably between the other couple of the electrodes, etc.

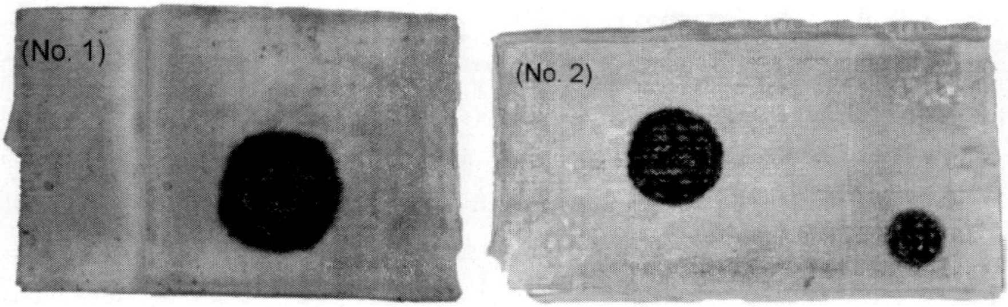

Figure 2. Oxide–nickel and cadmium electrodes of KSX-25 batteries (No. 1,2) after thermal runaway.

At the thermal runaway of the batteries KSX-25 in the elastic container, 297 liters of gas was collected (FIgure 1 (No. 1)).The gas analysis made with the aid of a gas analyzer showed that 95.6% of this gas was hydrogen (284 liters) and 4.4% oxygen (13 liters). The absolute error in the concentration percentage was 0.3-0.5. Besides, inside of the special reservoir of the coiled cooler, 152 ml of water was collected which corresponds to about 189 liters of steam released.

The oxygen release is connected with the electrolyte decomposition in accordance with the electrochemical reactions (3,4) in the course of the thermal runaway. Hence, during the thermal runaway, 20.9 ml electrolyte was decomposed. In the battery KSX-25, on average, 180 ml electrolyte is present. Besides, a part of the electrolyte can stay in electrodes' pores despite the very high temperature (more than 800°C) in the spot of the thermal runaway [19]. Besides, it should be observed that the amount of hydrogen released (284L(H_2)) is much more than its content in the entire electrolyte if we decompose it on hydrogen and oxygen (approximately 224L(H_2)).

In our earlier study [12], based on analysis of an energetic balance of a thermal runaway for a KSX-25 battery, it was shown that in the process of the thermal runaway, the battery emits energy at least 6 times larger than it obtains from the recharger. Moreover, in this work we estimated only the lower limit of the energy emitted. In fact, the energy released during the thermal runaway is much more. In the quoted assessment [12], only the energy was taken into account that was spent on the battery heating up and electrolytic solution evaporation but not the energy dissipated by the battery during the thermal runaway and not the energy spent on the meltdown of the battery plastic housing, and so on.

In this work we performed a calorimetric analysis of thermal runaway, Figure 1 (No. 1). Experimental research showed that in the result of thermal runaway, the energy equal to 5012 kJ (relative error 7%) was released from the battery. Meanwhile the battery receives energy equal to 33.7 kJ [12] (during the process thermal runaway) from the charging device. Usually during such a thermal runaway, all electrolyte evaporates and the plastic body of the battery melts down.

2.3. Discussion of the Mechanism of Thermal Runaway

Let's consider the standard mechanism of the thermal runaway initiation from the point of view of our experimental data received in this work and in earlier works [12-18].

According to the standard point of view [1, 20, 21], the thermal runaway in batteries has the following scenario. In a case of a battery, long recharging under constant voltage (or at its floating service), heating-up takes place, which results in the decreasing of a battery's internal resistance and an increase of recharging current, which in turn magnifies heating up, etc. The thermal runaway is the result of a positive feedback between the current and temperature of batteries during their charging under a constant voltage. Besides, as for hermetic batteries, there is a belief that an essential initial contribution to batteries heating-up are made by the exothermic reactions of an oxygen cycle.

The KSX-25 batteries investigated by us are not hermetic. They are the typical vented batteries with a closed-packed arrangement of the electrodes. This is why the oxygen cycle in these batteries is absent.

In our previous study (Section 2.1), it was shown that the probability for thermal runaway increases with an increase in temperature of the battery and an increase in charging voltage. This result corresponds to the standard mechanism of the thermal runaway [1, 20, 21] completely.

Indeed, both the increase of the constant charge voltage leading to the increase of the recharging current and the environmental temperature growth must considerably increase the initial temperature of the batteries and hence both these factors must increase the probability of the thermal runaway. But the other results do not correspond to this mechanism of the thermal runaway.

Firstly, according to the standard mechanism of the thermal runaway [1], at the same conditions, the thermal runaway must be initiated with an equal probability both in the new batteries and in the batteries with long operation lives. This follows from the fact that in the standard mechanism of the thermal runaway, the processes of the ageing of the batteries are ignored completely. But the direct experimental data (Table 4) just show that the thermal runaway probability depends a great deal on the battery operation life! This correlation is thoroughly confirmed (in private conversations) by technicians servicing alkaline batteries in airports. For at least thirty years of our studies of the thermal runaway phenomenon in alkaline batteries, we had no chance to observe the after effects of the thermal runaway in new batteries KSX-25 under actual operating conditions. Meanwhile, the after effects of the thermal runaways of the batteries of this type with the long operation lives working on real objects were studied and described by us many times. So the established correlation between the thermal runaway probability and the battery operation life obviously indicates that the ageing processes must be considered as integral part of thermal runaway mechanism.

Second, according to the standard mechanism of the thermal runaway, along with heating-up, an internal resistance of the whole system of a battery decreases. Hence, if some thermal runaway after effects are possible, they must be regularly spread on all surfaces of the electrodes. But a visual analysis of the thermal runaway after effects (Figure 2) shows that the thermal runaway takes place only in small areas having the regular shape of circles and in no way affects the rest of the electrode. Even the visual analysis of the burnt-through areas of the separators on the electrodes (Figure 2) indicates that in this area a very powerful exothermic process took place.

From the point of view of the standard mechanism of the thermal runaway, the shape of the separator burning-through areas looking like correct circles cannot be understood. A conclusion can be made that a local heating-up of the electrodes during the thermal runaway is connected with the non-uniformity of a charge current density on the electrodes' surfaces. This non-uniformity can be connected, for example, with a non-uniformity in the distribution of dendritic crystals in the separator on the electrodes' surfaces or with some other ageing processes of separators and electrodes, as a number of colleagues supposed in a private conversation. But no reason mentioned above can constantly result in the forming of powerful local current density in the shape of the correct circles. It is as if the thermal runaway is initiated from a point on a surface and then similarly to the fire growth process, it is uniformly distributed radially from this initial point.

Third, according to the standard mechanism, the thermal runaway takes place on expense of the energy, which the battery obtains from the recharger [1, 20, 21]. Calorimetric analysis (Section 2.2) showed that in the result of thermal runaway, the energy equal to 5012 kJ released from the battery. Meanwhile the battery receives energy equal to 33.7 kJ [12] from the charging device (during the process thermal runaway). This fact can be explained only from an assumption that the thermal runaway is accompanied by a powerful exothermic reaction. Moreover, this reaction must be the basis of the mechanism of the thermal runaway.

Fourth, the standard mechanism of the thermal runaway supposes that the thermal runaway is connected only with an intensification of a battery charge reaction (on expense of internal resistance decreasing) and possibly also of an electrolytic solution decomposition reaction as the thermal runaway takes place in the case of long recharging of the batteries. If this is the case, a hydrogen/oxygen ratio in the liberated gas must be 2 : 1. Nevertheless, the experiment [12, 14] shows that the liberated gas mixture contains more than 95% hydrogen.

Fifth, current I rises sharply when thermal runaway initiates, and consequently, the temperature of the electrodes increases sharply. In order to gain insight into the causes for the increase in current, we measured the voltage U at the battery terminals, Figure 1 (No. 1).

The charging current and voltage of the charging device are correlated by the following equation:

$$I = \frac{U - E}{r_{ex} + r_f}, \tag{5}$$

where I is the charge current, U is the voltage maintained by the charging device, E is the battery open circuit voltage (OCV), r_{ex} is the resistance of the connecting leads and shunt, r_f is the battery internal resistance. The resistance r_f is nonlinear; it takes into account not only ohmic resistance of electrolyte but also all other nonlinear processes, depending on current and influencing voltage on battery terminals U and in particular the activation processes on the electrode-electrolyte border.

Voltage at the battery terminals can be equated as follows:

$$U_t = E + I \cdot r_f . \tag{6}$$

According to the standard point of view [1, 20, 21], the cause of thermal runaway in the given mechanism is related to the decrease of internal resistance of the batteries due to their heating. Let us analyze the curves in Figure 1 (No. 1) for the change of current and voltage during thermal runaway.

During the charging of Ni–Cd batteries the reaction (1) takes place at the positive electrode and the reaction (2) takes place at the negative electrode. OCV corresponding to reactions (1,2) shall be E =1.26-1.44 V, depending upon the charge state of the electrodes and their type. The thermal runaway takes place due to long overcharging of batteries [1]. The only reaction at this moment can be the decomposition of water to hydrogen and oxygen (3,4).

Thus, in formulas (5,6) E shall be the OCV of a hydrogen–oxygen element, i.e., the difference between the standard potentials of hydrogen and oxygen electrodes in the given electrolyte, which is equal to 1.229 V [22].

Let us examine the initiation of the thermal runaway process. When it starts, the voltage at the battery terminals drops sharply to approximately 0.55 V (Figure 1 (No. 1)). Such a sharp drop in voltage, according to equation (6), cannot be explained by only the decrease in the internal resistance of the battery due to an increase in temperature. Actually, this voltage is even less than the OCV for the above reactions (1,2) and (3,4). This voltage drop can only be explained by the decrease in OCV during the process of thermal runaway. It is possible only if the thermal runaway process is related to the initiation of a new potent electrochemical reaction with a much lower difference (ΔE) in the standard potentials for the anode and cathode, as compared to reactions (1,2) and (3,4).

The generally accepted mechanism for thermal runaway [1, 20, 21] assumes that during the thermal runaway, just as before it starts, the electrochemical reaction for the decomposition of water as in equations (3,4) is underway, but much more intensively due to the decrease in the internal resistance of the battery. However, neither the reaction for the decomposition of water (3,4), nor the battery charging reaction (1,2) can proceed at a battery terminal voltage of 0.5-0.6 V during thermal runaway.

It is worth mentioning, that the parameter E decreases with the rise in battery temperature for reactions (3,4) according to the equation [23, 24]

$$E=1.229(1-0.0003(T-298)) \tag{7}$$

where T is the battery temperature.

However, it can reach a voltage of 0.5-0.6 V only when the temperature rises to 2000°K. Thermal runaway in the battery as seen from Figure 1, started at a temperature of approximately 373°K at the positive terminal of the battery. The temperature is much higher than 373°K [12], but much lower than 2000°K at the point of thermal runaway at the electrodes.

Hence, the analysis unequivocally shows that the thermal runaway is not related to the intensification of the reactions (1–4) due to the self-heating of the batteries, but from a new electrochemical reaction proceeding with a much lower difference in standard potentials ΔE between the anode and the cathode, than for reactions (1,2) and (3,4).

The given analysis confirms the conclusions made on the basis of thermal runaway energy balance in our earlier manuscript [12], that the thermal runaway is related to the initiation of a new potent exothermic reaction.

The sharp decrease in OCV (E) and battery resistance r_f during thermal runaway, according to equation (5), should produce a sharp growth in the charging current, as is observed in the experiment.

Further, the internal resistance of the battery sharply increases, due to evaporation of the electrolyte, and the charging current sharply drops to very small values as in Figure 1 (No. 1).

On the basis of all the obtained experimental data, the following mechanism of the thermal runaway in the nickel-cadmium batteries can be proposed [15].

As during the thermal runaway, a great mass of hydrogen is released from the batteries for 3-5 minutes (Section 2.2 and [12, 14]), the assumption can be made that the hydrogen is an active participant of the thermal runaway process and of the exothermic reaction accompanying this process. But the hydrogen is accumulated in the battery's electrodes only in proportion to their operation life [12, 13, 16]. Hence, if the assumption is correct that hydrogen is the active participant of the thermal runaway process, then the thermal runaway probability must grow so far as the hydrogen is accumulated, or so far as the battery spends its operational life, which exactly corresponds to the experimental data (Section 2.1). The other process of step-by-step ageing, leading the battery to the thermal runaway is the process of collection of dendritic crystals in the separator.

In a location area of a dendritic crystal, a distance is essentially decreased between the electrodes. This event results in the occurrence that in this area, the charge current density will be much more than in other locations of the electrodes. Hence this place will be heated up much more vigorously than the other locations of the electrodes.

When the cadmium electrode is heated up to 340°C, a remarkable hydrogen release is initiated [12], but even at temperatures at least one and a half times less, a minor hydrogen release is already initiated. This can be a reason for the exothermic reaction initiation of the thermal runaway. This reaction will heat up the electrodes even harder in the location of a dendritic crystal and in neighboring areas around it, which will lead to even greater hydrogen release and so on. Thus, the exothermic reaction will outspread from the location of the dendritic crystal radially and so it will burn the separator through in the shape of the correct circles (Figure 2). There are two factors which can prevent the outspreading of this exothermic reaction on the entire electrode surface. First is the gas lock, which is formed in the hole of the burnt-through separator and between the electrodes after the electrolytic solution evaporation. Second is hydrogen depletion in the electrodes in the area of the thermal runaway.

The dendritic crystals emerge in the separators in random locations; this can be an explanation of the absolutely random location of burn holes of the separators in different batteries (Figure 2).

2.4. Probability Investigation of Thermal Runaway in Nickel-Cadmium Batteries with Sintered, Pasted and Pressed Electrodes

For experimental studies, batteries were used with the following oxide–nickel electrodes: sintered (KSX-25, KSL-15, KSX-6, KSM-3.5), pasted (KM-3.5), and pressed (KL-24, KL-20, KL-10) [25, 26].

For the batteries' cycling process, a recharger was used which allowed installing one of a number of fixed charge voltages: 1.45; 1.67; 1.87; 2.2 V. The recharger allowed working on a constant basis with currents up to 300 A and on short-duration bases with currents up to 1000 A.

In order to obtain more voluminous statistics for a shorter period of time, the recharger was connected to a group of ten paralleled batteries in a hard metal clamp band. These batteries joining up in parallel was made with the help of two heavy duty metal runs, to which the batteries' positive and negative terminals were screwed on. In order that a thermal runaway developed in one battery had no influence on a probability of a thermal runaway development in neighboring batteries (which could happen because of their additional heat-up), between the batteries in the metal clamp band, the heat-insulating wood block fillers two centimeters thick were inserted.

The batteries were charged in succession under constant voltage values: 1.45; 1.67; 1.87; 2.2 V. The charging process lasted ten hours. Discharge was conducted according to the operation manual for each specific battery.

Since, in our previous studies [15], there was shown that the probability of a thermal runaway grows with an increase of operating life, in these new experiments, batteries were used with an operating life at least twice as large as their guaranteed service life, which supposedly would facilitate the thermal runaway process initiation. In a group of ten paralleled batteries, the batteries were grouped by those with the same operating life. Each group of batteries was cycled 80 times. This way, for each battery type and for each charge voltage value (1.45; 1.67; 1.87; 2.2 V), 10*80=800 charge-discharge cycles were conducted.

The results of the batteries' cycling are represented in Tables 5 and 6. In Tables 5 and 6, only charge voltages under which a thermal runaway took place are given; for those batteries, the operating life is specified.

Thus from 800 conducted charge-discharge cycles for each type of batteries in tough conditions of charging, i.e., under high charging voltage values, the thermal runaway was watched only in the case of the KSX-25 and KSL-15 batteries. Therefore, it is fair to say that the thermal runaway is quite a rare phenomenon.

Table 5. Results of batteries cycling with sintered electrodes

Batteries	**KSX-25**	**KSL-15**	**KSX-6**	**KSM-3.5**
Charge voltage (V)	1.87; 2.2	2.2	-	-
Period of operation (years)	6.5; 7	6.6	> 6.5	> 6.5
Number of charge-discharge cycles	800	800	800	800
Number of thermal runaways	2	1	0	0

Table 6. Results of batteries cycling with pressed and pasted electrodes

Batteries	**KL-24**	**KL-20**	**KL-10**	**KM-3.5**
Charge voltage (V)	-	-	-	-
Period of operation (years)	> 6.5	>6.5	> 6.5	> 6.5
Number of charge-discharge cycles	800	800	800	800
Number of thermal runaways	0	0	0	0

In all the cases of thermal runaway, the batteries in question were distinguished with operating lives more than six years in spite of their three-year guaranteed service period. These experimental results directly confirm the conclusions of the paper [15] that the probability of the thermal runaway initiation grows along with an increase of their operating life.

In all the observation cases of the thermal runaway, the batteries' charge was conducted under the voltage values 1.87 and 2.2 V, which significantly exceed the average voltage for such batteries operation in field conditions in float mode (1.35-1.5 B). So it may be deduced that the probability of the thermal runaway grows as batteries' charge voltage increases.

Not one of the batteries with pressed and pasted oxide–nickel electrodes (KL-24, KL-20, KL-10, KM-3.5) developed a thermal runaway (Table 6). This can be connected with the types of electrodes and separators used.

In the papers [15, 17], it was proved that a thermal runaway initiation is connected with dendrites' intergrowth through a separator. The dendrites cut the distance between electrodes drastically. Consequently, in dendrites' locations, electrodes will locally heat up to a large extent due to the fact that resistance in such spots would be much less, while an average charge density would be essentially higher than in other places in the vicinity of those spots on the electrodes. This is exactly a reason for a thermal runaway launching as judged by any thermal runaway mechanism; both the one described in the literature [1] and the one proposed in the papers [15, 17].

In the batteries with the pressed electrodes, the separators are used based on thick fabrics (battery KL-10). In the batteries KL-24, KL-20, the positive electrode plate is wrapped by an alkali-proof paper and placed into a housing made from a capron fabric.

As the process of intergrowth of cadmium dendrites depends strongly on a separator's thickness, structure and pores diameter, as the separator's thickness grows and pores diameter decreases, the process slows down considerably [20]. Even if a dendrite inter-grows in the batteries with thick separators, due to a large length, it will not be able to sustain the high charge density needed for an electrodes' significant local heating up.

Nevertheless, univocal asserting based on the conducted experimental studies (that in the batteries with the pressed electrodes a thermal runaway is impossible) it is undoubtedly premature as studying the four battery types does not provide sufficient statistical material. However, the analysis of the data on the thermal runaway as well as of these batteries' operation on various enterprises counts in favor of this presumption. In particular, an analysis of mining batteries' operation over the period of more than 40 years showed that a main reason of batteries 3KL-10 failures is an exfoliation of an active mass from an internal current-collecting net on an oxide–nickel electrode. The exfoliation occurs because of long-run batteries' overcharging. A gas released in a case of the overcharging (primarily on the current-collecting net because of its high conductivity) pulls away the active mass, step by step. A long-run overcharging can occur, for example, in a case that a coalminer did not come to work in his shift; in this case the battery stays connected to a charging stand not 16 hours as it is prescribed by an instruction, but instead three times longer. So the long-run overcharging of these batteries leads not to an immediate thermal runaway, but instead to the exfoliation of the electrodes active mass. In the batteries with the sintered electrodes, there cannot be such phenomena. So it is possible that the active mass exfoliation in the batteries with the pasted and pressed electrodes (with probability being higher than the same of the thermal runaway) does not let the thermal runaway process develop and in so doing it decreases the probability of thermal runaway in these batteries.

In any case, these experimental studies univocally show that the thermal runaway in the batteries with the pressed electrodes is either impossible or its probability is much lower than in the batteries with the sintered electrodes.

Also, based on the conducted experimental studies, one may not univocally assert that a reason of a thermal runaway's absence in these batteries is the separator thickness inasmuch as it is necessary to investigate thoroughly the mechanism of cadmium dendrites intergrowth through separators' various materials and also to study an impact of a structure and pores diameter of a separator on processes of the dendrites formation. Such investigations would allow deducing a dependence of dendrites growth speed on separator thickness and pores diameter. After that, it would be possible to determine, which separators should be considered as thin and which ones are thick enough from the point of view of the prognosis of a thermal runaway initiation. Such studies are beyond the scope of this paper. In our experiments, the batteries of low capacity did not develop the thermal runaway, no matter what kind of electrodes were installed on them: sintered electrodes (KSX-6, KSM-3.5) or pasted ones (KM-3.5) or pressed ones (KL-10). For more than 25 years of thermal runaway studying in alkali batteries, never have we coped with initiating a thermal runaway in batteries with a capacity less than 12Ah. Thus, a batteries' capacity significantly influences a thermal runaway's occurrence probability. From the point of view of the thermal runaway mechanism represented in the paper [17], this dependence can be explained by the influence of the following factors.

First, the higher a battery's capacity, the more its mass, and hence, the worse the heat-removal from internal electrodes. So (during charging at the same voltage), internal electrodes located inside batteries of a large capacity will be heated up more than internal electrodes of batteries of a small capacity. However, according to investigations described in the paper [15], with electrodes temperature growth, a thermal runaway occurrence probability grows as well.

Second, at the same charging voltage and hence at the same charging current density (as electrodes of the batteries in question are the same constructively), batteries of larger capacity are charged with large total current. In a case of a short circuit, through a dendrite, the large total current of charge will allow concentrating a larger local current in this spot and hence heating up this electrode more intensively than in batteries of a small capacity. Undoubtedly, both of these factors promote a thermal runaway initiation.

Thus, in this study, for the first time it was proved experimentally, that an occurrence probability of a thermal runaway in nickel-cadmium batteries with sintered electrodes falls down with a decrease of the batteries' capacity. However, based on the conducted experimental studies, a categorical affirmation about impossibility of thermal runaway in batteries of small capacity is, of course, premature as the studying of three battery types does not provide us with a sufficient statistical material. Besides, the conducted experimental research shows that in batteries KSX-25, KSX-6 and KSX-3.5, in equal measures, the accumulation processes are present (of hydrogen in electrodes [26] and of dendrites on cadmium electrode) necessary for a thermal runaway occurrence. Hence, according to the thermal runaway mechanism [17], in all the studied batteries, a thermal runaway process occurrence is possible. Although in batteries of a small capacity, in virtue of a better heat-removal from internal electrodes and lower charging total current, it is needed to use more severe charging conditions in order to obtain a thermal runaway process occurrence. For example, it is needed to use a higher charging voltage or a higher ambient temperature [15]. However, this assumption requires further theoretical and experimental research.

2.5. Probability Investigation of Thermal Runaway in Nickel-Cadmium Batteries with Pocket Electrodes

For the experimental studies, we chose batteries with pocket electrodes (KL-14, KL-28) and batteries produced by SAFT, specifically those with the pocket electrodes (SBLE 30 and SBM 22) [27].

The cycling was carried out for a group of ten parallel-connected batteries. This allowed us to obtain a large amount of statistical data in a relatively short period of time. Between the batteries we inserted heat-insulating gaskets made of wood 2 centimeters thick. Thus, thermal runaway arising in one battery could not influence the probability of occurrence of thermal runaway in the neighboring batteries due to their additional heating.

According to our previous studies [15], the probability of a thermal runaway increases with the growth of the following factors: charge voltage, batteries' operating life and ambient temperature. In connection with this, there were chosen batteries with an operating life of 7 years. The batteries were charged under the charging voltage 2.2 V for 10 hours and discharged by current down to 1 V in accordance with operational directives for each of those batteries. A cycling process was conducted in a thermal chamber under a temperature of 45°C. For each of the batteries groups there were 100 charging-discharging cycles fulfilled. So 10*100=1000 of the charging-discharging cycles were conducted for each type of batteries. The investigation results are represented in Table 7.

Table 7. Results of batteries cycling at an environmental temperature of 45°C

Batteries	**KL-14**	**KL-28**	**SBLE 30**	**SBM 22**
Charge voltage (V)	2.2	2.2	2.2	2.2
Period of operation (years)	7	7	7	7
Number of charge-discharge cycles	1000	1000	1000	1000
Number of thermal runaways	0	0	0	0

Therefore, despite the long operating life of those batteries and having conducting 1000 charging-discharging cycles for each type of battery (in tough conditions), a thermal runaway never took place, even in one of them. But in similar experiments conducted with the nickel-cadmium batteries with the sintered electrodes, (for example, for the batteries KSX-25), the thermal runaway occurred with a very high probability [12].

Consequently, at least this experiment set shows that the probability of a thermal runaway in nickel-cadmium batteries with pocket electrodes is much less than the same in nickel-cadmium batteries with sintered electrodes.

In favor of this affirmation, one can give other arguments, too. We have never managed to achieve a thermal runaway in the nickel-cadmium batteries with the pocket electrodes (during more than thirty years of this phenomenon investigation), while in the nickel-cadmium batteries with the sintered electrodes, the thermal runaway has been watched and studied by us repeatedly [12, 14-17]. Technicians maintaining nickel-cadmium batteries with the pocket electrodes univocally assert that there is no such thing as thermal runaway in batteries of this kind. We also have found no information on a thermal runaway possibility in the nickel-cadmium batteries with the pocket electrodes in scientific literature.

From the point of view of the classical mechanism of thermal runaway [1], the obtained experimental result is inexplicable as this mechanism does not take into consideration either the batteries' construction or their type of electrodes. Therefore, a thermal runaway should occur with the same probability for any nickel-cadmium battery with any type of electrodes, although the experimental results show that it's not like that (Table 7 and [27]).

From the point of view of the thermal runaway mechanism outlined in the paper [17], this experimental result is both natural and obvious.

The batteries with pocket electrodes are of two kinds: with free electrodes location (in this case as separators, there are used ebonite sticks, rubber cords, plastic grills; it concerns the batteries KL-28, SBLE 30, SBM 22, etc.) and with closed-packed arrangement of electrodes and micro-porous separators (battery KL-14).

In the first case, between the electrodes there are large distances through which dendrites usually cannot intergrow. Also dendrites intergrowth in these batteries is not facilitated either by a free electrolyte convection between the electrodes or by an absence of a mechanical support for dendrites in the form of separator material. In these batteries it is therefore impossible to formation stable dendrites to allow electrodes strong local heating up for an exothermic reaction launching; and hence there cannot be any thermal runaway.

In the case of the batteries with a closed-pack arrangement of electrodes and thin separators, the electrodes strong local heating up is impossible too. Hence, the launching of an exothermic reaction is also impossible. Indeed, even if a dendrite intergrows between the electrodes of this construction, it will close the chain onto the metallic pocket of the opposite electrode and simply burn down, not causing any essential local heating up by virtue of a high conductivity of the pockets metal.

So in the batteries with the pocket electrodes, it is impossible that there would occur a powerful local electrodes heating up connected with dendrites intergrowth, while namely, this phenomenon is the reason for thermal runaway initiation [17]. That is why in these batteries, the thermal runaway is totally impossible. But this statement requires further experimental and theoretical investigations. Nevertheless, the experimental studies conducted, univocally show that the probability of the thermal runaway in the nickel-cadmium batteries with the pocket electrodes is many times less than the thermal runaway probability in the batteries with the sintered electrodes. This result can be of a great practical importance for the development of new types of nickel-cadmium batteries with good thermal runaway resistance.

3. Thermal Decomposition of Alkaline Battery Electrodes

3.1. Thermal Decomposition of Sintered Electrodes

For the experiments conducted, the batteries KSL-15 were selected (with sintered electrodes by capacity 15 Ah) with an operating life of from 0 to 7.5 years and the batteries KSX-25 (with sintered electrodes by capacity 25 Ah) with operating lives from 0 to 7.5 years [13, 16, 28-30].

The amount of hydrogen accumulated in the electrodes was found by their heating in the metal thermal chamber.

The experimental stand for the process investigation of the gas evolution from battery electrodes at their heating up is shown in Figure 3. It was represented by a metal thermal chamber in a form of a tube 1.8 m long and 2 cm in diameter. A sealed end of the tube was placed into a muffle furnace, while into the other end a rubber plug with a pipe for gas withdrawal was inserted.

For the convenience of electrodes' extraction from the thermal chamber in their coiled state, the electrodes were placed inside a metal cartridge and then the cartridge was placed into the thermal chamber. In order to reduce a heat exchange in the thermal chamber, a round porous ceramic plug 20 cm long was placed above the cartridge.

Gas escaping from the thermal chamber 7 as a result of electrodes heating up, passed through a standard coil 9, cooled down and got into an intake chamber 2. The intake chamber 2 was represented by a hermetically sealed glass reservoir 5 liters in volume filled with water. At the bottom of this reservoir, three tubes were inserted through a rubber plug. Via the middle tube (Figure 3), the gas from the thermal chamber got to the upper part of the reservoir. The tube on the right (Figure 3) looked like a leg of a manometer 1 and was used for pressure control of the gas accumulated above water. The tube on the left (Figure 3) was intended for excess water removal from the reservoir.

When the gas from the thermal chamber 7 got into the intake chamber 2, the gas pressure above water increased and this fact was fixed by the manometer 1. In order to establish the pressure inside the intake chamber being equal to the atmospheric pressure, a portion of the water was removed from the intake chamber into retort 4. The water amount in the retort determined the amount of gas released from the electrodes in the thermal chamber during the studied period of time [12, 16, 28].

A decomposition of every electrode was conducted at the temperature 800°C. This temperature was chosen based on the following considerations. In our preliminary experiments, it was found that for a cadmium electrode, a remarkable gas release starts at 340°C, while for oxide-nickel one at 660°C. For both oxide-nickel and cadmium electrodes, an essential gas release starts at temperatures more than 740°C. Thus the temperature 800°C was chosen as an optimal one for the thermal decomposition for both cadmium and oxide-nickel electrodes.

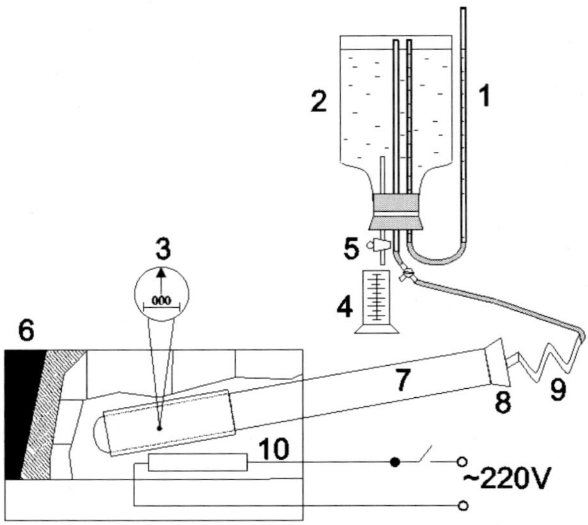

Figure 3. The experimental set-up to study the process of gas release from the heated electrodes. (1) Manometer, (2) intake chamber accumulating gas, (3) thermocouple, (4) retort, (5) tap, (6) muffle furnace, (7) metal thermal chamber, (8) rubber plug with a tube for gas extraction, (9) standard coil and (10) heater.

The thermal decomposition process stopped when a daily gas release stayed less than 4 mL/g (of an electrode). For example, for a battery KSL-15 with a long operating life, the thermal decomposition took on average 12 days for a cadmium electrode and 14 days for an oxide-nickel one, on the basis of an 11 hour-long working day. Daily, from a cadmium electrode, not less than 3.5 L released on first day and up to 100-110 mL gas on last day, while for an oxide-nickel electrode, the amount made not less than 5.5 L on the first day and up to 100-110 mL on the last day.

A quantitative determination of the released gas composition was made with aid of the gas analyzer VOG-2M.

The results of the experimental investigations for the battery KSL-15 are represented in the Table 8, while for the battery KSX-25, they are in

Table 9. To the thermal decomposition, one electrode was subjected, which preliminary was rolled up in the form of a tube and placed into the cartridge.

At the thermal decomposition of both oxide-nickel and cadmium electrodes, from new batteries KSL-15, the gas mixture release is in the amount about 180-192 mL. This volume falls within the device's accuracy limits. The point is that at an initial moment at electrodes heating up in the thermal chamber, the air enters into the measuring container because of air thermal expansion in the thermal chamber itself. Then, with the thermal chamber heating up from the electrodes, the gas starts releasing and together with the hot air it enters into the measuring container. So it was very difficult to determine the moment the gas was released from the electrodes and into the measuring container.

This is why a set of experiments was conducted on the heating up of an empty thermal chamber to various temperature values. When the empty thermal chamber was heated up to the temperature 800°C, air in amount of about 180-190 mL entered the measuring container. So in connection with the air heating up in the thermal chamber, we let out the first 210 mL of gas into the atmosphere, while the remaining gas was collected and analyzed.

An analysis of the gas released from the electrodes of both types with aid of the gas analyzer VOG-2M showed that the released gas consists only of hydrogen. The absolute error in the percentage concentrations is 0.3-0.5.

Table 8. Dependence of average volume of gas released (from one electrode) on operating lives of batteries KSL-15 at thermal decomposition of oxide-nickel and cadmium electrodes

No. of the battery	**1**	**2**	**3**	**4**	**5**	**6**	**7**	**8**	**9**
Period of operation (years)	New	1	1.5	3.5	4	5	5.5	6	7.5
Amount of gas released (Ni) (L)	0	13	16.7	26.3	29.9	32	31.5	31.6	31.9
Amount of gas released (Cd) (L)	0	11	12.8	16.1	18.7	20	19.5	19.5	19.8

Table 9. Dependence of average volume of gas released (from one electrode) on operating lives of batteries KSX-25 at thermal decomposition of oxide-nickel and cadmium electrodes

No. of the battery	**1**	**2**	**3**	**4**	**5**	**6**	**7**	**8**	**9**
Period of operation (years)	New	1	1.6	3.3	4.5	5	5.5	6.1	7.5
Amount of gas released (Ni) (L)	0	14	19	27	33	36	35.8	36	35.9
Amount of gas released (Cd) (L)	0	10	13	16	18	20	20	20.1	19.9

The results of the conducted experiments (Tables 8 and 9) show clearly that the hydrogen is accumulated in oxide-nickel and cadmium electrodes in proportion to the batteries operation duration as the hydrogen amount in the oxide-nickel and cadmium electrodes grows as the batteries' operating life increases. It is notable that new oxide-nickel and cadmium electrodes do not contain any hydrogen at all.

At a rather short operating life, the content of hydrogen in electrodes varies very much. Apparently, it is connected with different conditions of various batteries' operation. At long operating lives (more than five years), the hydrogen content in the electrodes is approximately the same.

In batteries with a long operating life, a really great amount of hydrogen is accumulated. So, for example, the battery KSL-15 is comprised of 5 cadmium and 6 oxide-nickel electrodes. Hence, in one battery KSL-15 (No. 6, Table 8), the hydrogen amounts to approximately 292 L. The battery KSX-25 is comprised of 14 cadmium and 15 oxide-nickel electrodes. Hence in one battery KSX-25 (No. 6, Table 9), the hydrogen content makes approximately 820 L. These values correspond to values obtained previously for other types of nickel-cadmium batteries [13, 16, 28-30].

From the following calculations, the fact is seen that in the course of the batteries' operation, this amount of hydrogen accumulation is possible:

According to the technical operations manual of the batteries KSL-15, they are charged with the current 3.8A for 6 hours. Therefore, these batteries are recharged 1.52 times more as compared with their nominal capacity. The recharging is needed for batteries' complete charging. So at the charge, almost 7.8Ah is spent on the electrolyte decomposition with the evolution of hydrogen and oxygen.

So, in one act of charging the battery KSL-15, up to 3 liters hydrogen and up to 1.5 liters oxygen are released. Thus, the 292 liters hydrogen in the KSL-15 electrodes can be obtained after 98 charge-discharge cycles. Close to the end of its operating life, a battery KSL-15 has been subjected to a number of charge-discharge cycles being ten times greater. This is why, theoretically, the battery can accumulate the found amount of hydrogen.

According to the operations manual of the KSX-25 battery, it is charged in two stages: first by a current of 10A over a period of 3 h, and second by a current of 5A over a period of 2 h. Consequently, these batteries overcharge by a factor of 1.6 as compared to their nominal capacity. Overcharge is required for the complete charging of the batteries.

Thus, at the charging of a battery KSX-25, almost 15 Ah is spent on the electrolyte decomposition and the release of hydrogen and oxygen. So at one charging cycle of a battery KSX-25, up to 6 liters hydrogen and up to 3 liters oxygen is released. Thus, the 820 liters hydrogen found in the electrodes of the batteries KSX-25 can be obtained in 137 charge-discharge cycles.

During five years of operation, the batteries KSX-25 are subjected to a ten times greater number of charge-discharge cycles. That is why, in principle, they are able to accumulate the amount of hydrogen found.

At the charging of nickel-cadmium batteries with direct current, the water decomposition takes place with the evolution of hydrogen and oxygen. Nevertheless, the experiments (Tables 8 and 9) show that no oxygen is accumulated in the electrodes, while hydrogen is. And moreover, the latter is accumulated in very big amounts both in oxide-nickel and cadmium electrodes. This is connected with the fact that the hydrogen possesses a very high diffusion permeability. For example, at the temperature 20°C, a diffusion coefficient of the hydrogen in nickel is approximately 10^{10} times higher than the diffusion coefficient of the nitrogen or oxygen [31]. In this way, at electrolyte decomposition onto hydrogen and oxygen, only the hydrogen penetrates into the electrodes and is accumulated inside of them, while the oxygen exits into the atmosphere.

In the paper [16], it was shown that hydrogen accumulates in a sintered nickel matrix of an oxide-nickel electrode.

Now let us evaluate specific parameters of a sintered nickel matrix of an oxide-nickel electrode as a hydrogen accumulator. According to the experimental data (No. 6, Table 9), in a sintered nickel matrix approximately 36 liters hydrogen is accumulated during the long operating life of the batteries KSX-25. The weight of an oxide-nickel electrode is 24 grams, while the weight of the active substance of it is 8 grams. So the gravimetric capacity of the oxide-nickel electrode as a hydrogen absorber is equal to 13.4 wt% and the gravimetric capacity of a sintered nickel matrix of an oxide-nickel electrode as a hydrogen accumulator is equal to 20.1 wt%. The obtained result exceeds previously obtained results for any reversible metal hydrides including magnesium hydride or complex hydrides by 3 times [31, 32]. Considering that the physical dimensions of an oxide-nickel electrode of a KSX-25 battery is equal to $7.3 \times 13.6 \times 0.081$ cm, we obtain that the volumetric capacity of an oxide-nickel electrode as a hydrogen absorber equals to 400 kg m^{-3}. From the physical point of view, the size of an oxide-nickel electrode of the battery KSX-25 is equal to the same of a sintered nickel matrix. Consequently, also their volumetric capacities are equal (400 kg m^{-3}) [16, 28]. This result exceeds the earlier obtained results for any reversible metal hydrides by more than 3 times [31].

As for causes why specific capacity parameters of the oxide-nickel electrodes are so high, they are not quite clear yet. For those causes clarification, further experimental and theoretical research is needed. However, it should be noted that in an oxide-nickel electrode, there is present a lot of hydrogen accumulation promoting factors that are under extensive investigation now.

First, the oxide-nickel electrodes in the batteries KSX-25 are made of finely divided nickel powder by way of its sintering. A crystal structure of these powders is destroyed to a large extent. As is well known, dislocations and other disorders of a crystal structure are centers of adsorption and traps for hydrogen as they lower hydrogen atom energy as compared to their placement in a normal interstitial site. This promotes the hydrogen penetration in the depth of the metal. Thus, defects of a metal crystal structure result in a drastic increase of hydrogen miscibility in it. With this purpose in mind, in the frame of the modern methods of metal hydrides' preparation for hydrogen storage systems, metals are ground down in ball-mills [33, 34].

Second, the oxides of the transition metals are catalysts of the hydrogen accumulation [35, 36]. However in the oxide-nickel electrodes, the role of an active substance is played exactly by an oxide of transition metals (nickel oxide).

Third, in the battery KSX-25, electrodes are packed tightly. At the charging of the batteries, the hydrogen releases on the cadmium electrode (2). However, due to the tight packing of the electrodes, it is able to penetrate inside the pores of the electrodes, both oxide-nickel and cadmium. The accumulated-in-metal hydrogen concentration C_H is connected with the hydrogen pressure P (see the following formula) [37, 38]

$$C = K\sqrt{P}\,. \tag{8}$$

The capillary pressure in the pores is determined by the formula:

$$P=A/R. \tag{9}$$

where A = 0.2 μm MPa is the proportionality constant for the given electrolyte, R is a pore radius.

In a porous electrode mainly, the pores have the average radius of the order of several dozens of microns. However, the finely divided powder being used for manufacturing of the sintered matrix of the electrodes has micro fissures with sizes from several hundreds of an angstrom to the size of crystalline lattice of the metal. Thus, hydrogen accumulating in the micro fissures as the result of the electrolyte decomposition can have the pressure value up to 100 MPa and higher [12]. Hence, most probably, the oxide-nickel electrode accumulates the hydrogen not electrochemically but instead at the expense of the high capillary pressure.

Fourth, the process of hydrogen accumulation in the batteries' electrodes runs through their whole operating life. In the batteries in question, this period is more than five years. In the timeframe of the existing methods, the process of obtaining hydrides lasts from some minutes up to several hours. Our analysis of papers devoted to studying hydrides for the last 20 odd years showed that most probably, experiments have been absent on the obtaining of hydrides of transition metals at high hydrogen external pressures, acting for a very long time. However, the time of external pressure exposure is an essential factor in any process of penetration of one substance into another. It is especially so, if the penetration process is very slow, as it is in our case.

Most probably, it is exactly the favorable combination of all of the above mentioned factors in alkaline batteries, that allows the accumulation of hydrogen in their electrodes in these huge amounts (Tables 8 and 9).

3.2. Thermal Decomposition of Pocket Electrodes

For an estimation of the amount of hydrogen stored in electrodes of the batteries FL-250, they were exposed to thermal decomposition in a thermal chamber (Figure 3) [39]. In each experiment, one pocket from an electrode was exposed to thermal decomposition. This was determined by the thermal chamber size. Each electrode was decomposed at 600°C. The reasons for choosing this temperature were as follows. In preliminary experiments, it was determined that the visible release of gas starts at about 200°C. A big release of gas starts at temperatures in excess of 600°C for both the oxide–nickel and iron electrodes. Thus, the temperature of 600°C was chosen as an optimum value for thermal decomposition of both the iron and oxide-nickel electrodes.

For the experiments in a random way, five batteries FL-250 were selected with different service lives. All batteries were operated under the same conditions. In each battery, three pockets of both oxide-nickel and iron electrodes were exposed to thermal decomposition. The thermal decomposition of each pocket lasted on average 9 to 12 days, 11 to 13 hours per day. In the experiments, within a day, from a pocket, an average volume of 3 L hydrogen was released from an iron electrode and 4.5 L hydrogen from an oxide-nickel electrode. The thermal decomposition process stopped when the hydrogen release became 80 ml or less. The experimental results are represented in Table 10. In the Table 10, there are shown the average values of the gas released for the three pockets from each battery.

An analysis of the released gas in the experiment using VOG-2M gas-analyzer showed that it consisted only of hydrogen. The absolute error in the percentage concentrations is 0.3-0.5.

Table 10 (batteries No. 4 and 5) demonstrates that the electrodes of the FL-250 batteries with long service life contain a huge amount of hydrogen. The data of Table 10 refers to one pocket of an electrode. In the battery FL-250, there are 11 iron and 10 oxide-nickel electrodes. Notably, an oxide-nickel electrode consists of 16 pockets, while the iron one has 13 pockets. So in one battery FL-250 there are approximately 13373 L hydrogen, made up of 8640 L in oxide-nickel electrodes and 4733 L in iron electrodes.

Table 10. Hydrogen content in pockets of the oxide-nickel and iron electrodes (batteries FL-250) with different service life[a]

No. of the battery	**1**	**2**	**3**	**4**	**5**
Period of operation (years)	New	3.2	4.5	6	10.6
Amount of gas released (Ni) (L)	0	33.4	50.2	54.0	53.9
Amount of gas released (Fe) (L)	0	11.6	25.8	33.1	33.2

[a]The relative error in the data in Table 10 is 3-5%.

The batteries FL-250 are charged with the current 65A for 7 hours. Consequently, these batteries are overcharged by a factor of 1.8 as compared to their nominal capacity. The overcharge is required for complete charging of the batteries. Direct experimental measuring showed that at charging, the battery FL-250 produces on average, 80 L hydrogen and 40 L oxygen. Thus, the discovered 13373 L of hydrogen may have accumulated during 168 charging-discharging cycles. But by the end of its life span, a FL-250 battery is likely to have gone through ten times more charging-discharging cycles. Therefore, it can – in principle – accumulate the amount of hydrogen we found.

Hydrogen is absent in the electrodes of new nickel-iron batteries, but during their service life, the amount of hydrogen inside the electrodes increases (Table 10, batteries No. 1-5). Our many-year experimental research showed that the hydrogen volume absorbed by electrodes stops increasing after 5 or 6 years of their service lives depending on the intensity of the batteries operation (Table 10).

Now let us evaluate the specific parameters of an oxide-nickel electrode as a hydrogen absorber. An oxide-nickel pocket accumulates about 54 L hydrogen during the long service life of the battery FL-250 (Table 10, batteries No. 4, 5). The active mass weight of a pocket before hydrogen desorption is 22 grams. So the gravimetric capacity of the active mass of the oxide-nickel electrode as a hydrogen absorber is equal to 22.0 wt%. The active mass of the oxide-nickel electrode consists of nickel hydroxide and graphite. The obtained result exceeds the earlier obtained results for graphite (using traditional methods) by five times [40], and the best results obtained for nanostructured carbon by 2 times [31, 41–46].

Considering that the physical dimensions of an oxide-nickel pocket of a FL-250 battery is equal to $14.9 \times 1.3 \times 0.55$ cm, we obtain that the volumetric capacity of active mass of an oxide-nickel electrode as a hydrogen absorber is equal to 444 kg m^{-3}. This result significantly exceeds the earlier obtained results for nanostructured carbon [31, 47].

Also, with due consideration that the weight of the active mass of the iron electrode is equal to 17.49 g and its sizes are $14.9 \times 1.69 \times 0.29$ cm, we obtain that the gravimetric and volumetric capacities of the iron electrode's active mass are equal to 16.9 wt% and 406 kg m^{-3} respectively. These values are close to the values obtained by us earlier for a sintered oxide-nickel electrode [16].

Now, let us give consideration to reasons of such high specific capacities of hydrogen storage in an active mass of oxide-nickel electrode. First of all, it must be noted that in an oxide-nickel electrode there are many factors, which contribute to hydrogen accumulation. At present, these factors are under intensive research.

First, in the paper [48] it was shown that in a case of pure carbon nanostructures, it is impossible to reach a high gravimetric capacity of hydrogen storage. An increase of the gravimetric capacity of hydrogen storage is possible only by way of doping of carbon nanostructures [49] or by changing of their structure, particularly by decreasing the size of their pores [50, 51]. In the paper [50] based on the quantum-thermodynamic model, there was demonstrated theoretically that by decreasing the pores' size, the gravimetric capacity of carbon nanostructures grow. Notably, the pores optimal diameter was found to be approximately equal to 0.7 nm. The gravimetric capacity growth along with pores' size decrease is explained by the growth of two factors' values. The first factor is growth of the hydrogen binding energy since the overlap of the interaction potential fields from the interior pore surface will enhance the carbon–hydrogen interaction. Second, smaller pores have higher specific surface area (SSA) for a given pore volume. These theoretical conclusions were confirmed experimentally on a great deal of carbon nanostructures [51, 52]. Now, if we perform the same calculations as in the paper [50] (keeping in mind that supposedly it is not hydrogen molecules that penetrate but instead hydrogen atoms), then for pores' optimal size we'll obtain the diameter 0.3-0.35 nm. This size is equal to the interplanar distance in the graphite crystal lattice. So for atomic hydrogen, an optimal condition is invasion into the graphite interplanar space, and the graphite powder is the optimal medium for the atomic hydrogen storage. Besides, the impact of the above highlighted two factors (promoting increase of material gravimetric capacity) grows drastically in a case of the atomic hydrogen penetration into the graphite. First, the hydrogen binding energy increases tens times as compared to the hydrogen binding energy on the graphite surface (or any other carbon nanostructure, even with taking doping by metal atoms into consideration), which is shown directly in experiments in the papers [53-

55]. Second, for the hydrogen atoms inside of graphite, a specific surface area (SSA) will be equal to surface of all carbon layers in the graphite, and herewith this SSA exceeds the SSA of any other carbon nanostructure by many times. So if the aim is to obtain an atomic hydrogen absorption, then based on all the performed theoretical and experimental research [50-52], one can affirm that an optimal medium for atomic hydrogen penetration is the graphite powder (or any other carbon nanostructure with the same density of carbonic layers). Moreover, in the case of atomic hydrogen penetration inside of the graphite, the gravimetric capacity of hydrogen storage must grow many times, which is what is seen in our experiments.

Third, in a great number of works, we see that the doping of carbon nanostructures with alkaline or transition metals leads to considerable growth of their gravimetric capacity [49, 56, 57]. This phenomenon is connected both with a general energy increase of binding between hydrogen and carbon nanostructures and with hydrogen molecules dissociation on metal surfaces and their further penetration (already in atomic form) into the carbon nanostructures [57], and this is especially important for graphite powder. It should be noted that in the oxide-nickel electrode of the battery FL-250, the graphite powder is just doped with nickel.

Fourth, in the case of metal hydrides and graphite, a considerable growth of the gravimetric capacity is obtained in the case of use of finely dispersed powders (they can be prepared by milling in ball mills) [33, 57-58]. Any imperfections of crystalline structures (particularly dislocations) are traps for hydrogen, as they decrease the energy of hydrogen atom as compared to their location in normal interstice. Besides they are the centers of hydrogen absorption, and also contribute to hydrogen penetration into the graphite's depth. Hence, imperfections of the graphite crystalline structure cause a sharp rise of hydrogen miscibility in it.

Besides, in micro-defects of graphite, there can be developed a very high capillary pressure of hydrogen [12, 28], which also will promote hydrogen diffusion into graphite. It should be noted that as for the oxide-nickel electrode of the battery FL-250, namely the finely dispersed powder, graphite is used.

Fifth, after six-year-operation, electrodes in batteries FL-250 are very densely stacked. Thus, hydrogen evolved at the iron electrode during the charging of the battery may penetrate into the pores of both oxide-nickel and iron electrodes. Hence, in all probability, the oxide-nickel electrode accumulates hydrogen not electrochemically, but due to the high capillary pressure. It was shown in paper [12] that capillary pressure in the battery's electrodes can reach 100 MPa and more.

Sixth, for the increasing gravimetric capacity of carbon nanostructures, various catalytic agents are used, in particular KOH [51, 59, 60]. In the paper [59], it is shown that via KOH activation, the hydrogen capacity of mesoporous carbon is enhanced one and a half times. It should be noted that in the battery FL-250, the KOH solution is used as an electrolyte.

Seventh, it is a well-known fact that oxides of transition metals act as catalysts of hydrogen accumulation in metals [35, 36]. As for carbon nanostructures, this question has not been studied enough up until now. We believe that the transition metals oxides promote hydrogen molecules dissociation, releasing atoms. This is very important for hydrogen atoms diffusion into metals and graphite but it is not important for superficial adsorption of hydrogen molecules. We believe that is why this question has not been investigated enough. So we conducted a number of single experiments for studying catalytic characteristics of nickel hydroxide used in the oxide-nickel electrodes of batteries FL-250. With this purpose in mind, in our first experiment we charged a new oxide-nickel electrode (free of hydrogen) of battery FL-250 in dense stacking condition with two iron electrodes for 750 hours with current density 16 mA cm^{-2}. This current density corresponds to the common current density, which the battery FL-250 charges with. The long-time recharge is needed for electrolyte decomposition into hydrogen and oxygen and for hydrogen storage in electrodes. The result was that the measured gravimetric capacity of active mass of oxide-nickel electrode made 0.5 wt%. In the second experiment, the active mass of oxide-nickel electrode was replaced with the graphite powder used in this electrode as a filler. The charging of this electrode was made in the same conditions as in the first experiment. The result was that the measured gravimetric capacity of the graphite powder made only 0.03

wt%. It means that the nickel hydroxide used in the oxide-nickel electrode of the battery FL-250 possesses a very powerful catalytic action for the process of hydrogen penetration into graphite powder.

Finally, the process of hydrogen accumulation in the batteries' electrodes takes place all through their service life. This is more than six years in the present study. During the charging process of the battery FL-250, its electrolyte decomposes into hydrogen and oxygen. Hydrogen is characterized by a very high diffusion permeability. For example, at the temperature 20°C the diffusivity of hydrogen in nickel is approximately 10^{10} times higher than the diffusivity of nitrogen or oxygen [31]. These features explain why, upon dissociation of an electrolyte into hydrogen and oxygen in a battery, only hydrogen accumulates in the electrodes, while oxygen escapes to the atmosphere.

In the modern methods of hydrogenation, the hydrogen accumulation process in carbon materials can take from several minutes to several hours. For the last 20 years, analysis of scientific research works conducted on hydrogen accumulation in carbon materials showed that there were probably no experiments on hydrogen accumulation in carbon materials under high external hydrogen pressures (which can exist at graphite micropores) during a long enough period of time (several years). However, the impact of the external pressure as well as the duration of the process are the important factors in any process of penetration of one substance into another. It is especially so when the penetration process is as extremely slow and difficult as in our case.

That was the favorable combination of all these factors, in all probability, that allowed storing such a great amount of hydrogen in the electrodes of the batteries FL-250.

We would also like to highlight that we just recently discovered the phenomenon of large volumes of hydrogen stored in the electrodes of Ni-Cd and Ni-Fe batteries in the course of their operation [12]. This phenomenon is quite unusual. It allows for accumulating hydrogen in metal hydrides and carbon materials in volumes considerably largerthan in all known materials obtained with use of traditional methods.

Undoubtedly, investigation of this complicated and poorly studied phenomenon requires further research, both theoretical and experimental.

3.3. Processes of Hydrogen Release Relaxation at Thermal Decomposition of Electrodes of Nickel-Cadmium Batteries

In this section, there is studied and modeled, the dynamics of hydrogen release from electrodes of nickel-cadmium batteries at their thermal decomposition [61]. The electrodes' thermal decomposition was performed in the thermal chamber (Figure 3). For research, the nickel-cadmium batteries KSX-25 were used by capacity 25Ah with sintered electrodes.

According to our previous studies [16, 28] and (Section 3.1), a top amount of hydrogen is accumulated in electrodes of batteries of the kind after five years of operation. That is why in the experiments, batteries were used with their operation period equal to seven years. The electrodes' thermal decomposition took place at temperature 800°C, as an essential hydrogen release would start for oxide-nickel and cadmium electrodes at temperature values more than 740°C. The process of the thermal decomposition was stopped when the amount of gas released during a 24 h period was less than 100 ml.

A gas release velocity was determined on the basis of an escaped gas volume during 5 minutes. Measuring of the gas release velocity was performed every hour.

The thermal decomposition of the cadmium electrodes was performed during 72 to 84 hours, while the same was done for the oxide-nickel electrodes during 130 to 140 hours. The process used to be performed for 10 to 13 hours per day. Then for a night, the process used to be interrupted for 11 or 14 hours and restored next day. Notably, the day after an installation's turning-on, the gas release velocity was always much higher than at the moment of an installation shutdown.

The results of the dynamics measuring of the gas release from the electrodes at their thermal decomposition for the first two days & nights are represented in Figures 4 and 5. The total amount of gas released from an oxide-nickel and a cadmium electrode is represented in Table 11. An analysis of the gas released – performed with aid of the gas analyzer VOG-2M – showed that the gas consisted only of hydrogen.

In paper [16], it was shown that the hydrogen is accumulated only in a ceramic-metal matrix of an electrode. This feature of the gas release (Figures 4 and 5) testifies that in an electrode, hydrogen is distributed on the entire volume of a ceramic-metal matrix of an electrode. Then as a result of the gas release in boundary layers of the ceramic-metal matrix of the electrode, a hydrogen concentration decreases step-by-step, so the gas release velocity drops. During interruptions of the experiment, due to diffusion processes, a leveling of hydrogen concentration takes place on the entire volume of the porous matrix. The next day, at the moment of the installation turning-on, the concentration of the hydrogen in surface layers of the electrode ceramic-metal matrix will be higher than at the moment of the installation shutdown. This can be an explanation of the considerable increase of gas release velocity at the moment of the repeated installation turning-on. Notably, this pattern of gas release is possible only in the case that a hydrogen diffusion velocity in the electrode ceramic-metal matrix is very low, while its amount is very large.

Table 11. Content of hydrogen in oxide-nickel and cadmium electrodes of a battery KSX-25

Type of electrode	**Ni**	**Cd**
Period of operation (years)	7	7
Electrode size (cm)	7.3×13.6 × 0.081	7.1×13.6×0.065
Amount of hydrogen released (L) [a]	35.9	18

[a] The relative error in the data in Table 11 is 5–7%.

Let us evaluate parameters of the hydrogen diffusion in the electrodes. As the process of the gas release at the thermal decomposition of electrodes is conditioned by diffusion processes, changes of a hydrogen flow from the electrodes in the course of time must be described by the Cottrell's equation [62]:

$$j_d = \frac{C_H S D}{\sqrt{\pi D t}}, \quad (10)$$

where j_d – hydrogen atoms flow from an electrode, C_H – per-unit-volume concentration of hydrogen atoms in the electrode, S – electrode surface area, D - diffusion coefficient of hydrogen atoms in a ceramic-metal matrix of an electrode.

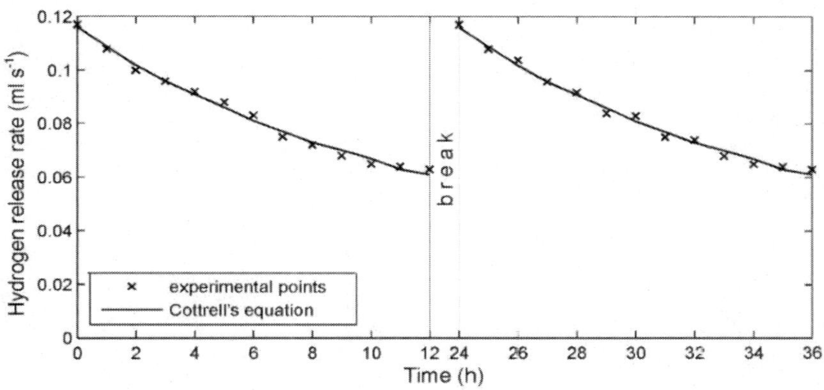

Figure 4. Process of hydrogen release relaxation at thermal decomposition of an oxide-nickel electrode of a battery KSX-25 at the temperature 800°C.

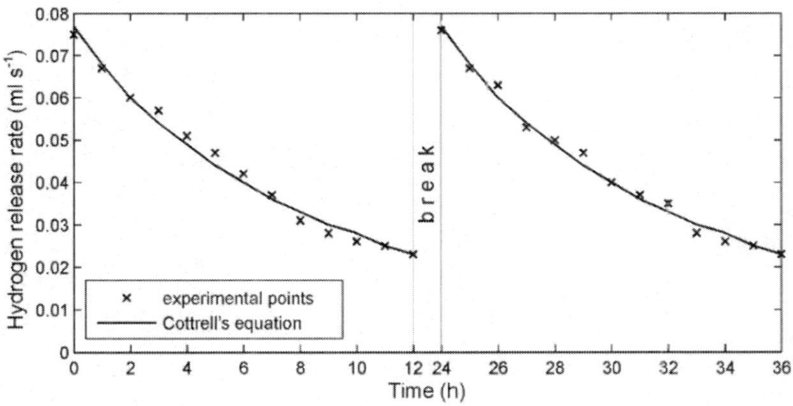

Figure 5. Process of hydrogen release relaxation at thermal decomposition of a cadmium electrode of a battery KSX-25 at the temperature 800°C.

Along with it, a diffusion boundary layer thickness in the ceramic-metal matrix of the electrode is set by the equation:

$$\delta = \sqrt{\pi D t}\ . \tag{11}$$

In our experiments (Figures 4 and 5), we started measuring the hydrogen flow parameters not from the moment of time, when the thickness of the diffusion boundary layer is equal to zero (as in the equations (10,11)), but instead from the moment when the temperature in the thermal chamber would reach 800°C. So at the moment the experiment starts (Figures 4 and 5), already the diffusion boundary layer had some thickness. In order to take this factor into account, we shall rewrite Cottrell's equation in the following form:

$$j_d = \frac{C_H S D}{\sqrt{\pi D(t + t0)}}, \tag{12}$$

$$\delta = \sqrt{\pi D(t + t0)}\ . \tag{13}$$

where *t0* – empiric parameter needed for taking into account the diffusion layer thickness at the moment the experiment starts.

Besides, a reverse flow of hydrogen from a thermal chamber into an electrode is possible. As in a thermal chamber (in a course of a thermal decomposition of an electrode), the constant temperature (T=800°C) and the constant pressure (equal to the atmospheric pressure) [12] are maintained, hence the reverse flow j_b must also be constant. So the equation for the hydrogen cumulative flow from an electrode will take the form:

$$j_s = \frac{C_H S D}{\sqrt{\pi D(t + t0)}} - j_b \tag{14}$$

For the purpose of conveniently comparing Cottrell's equation (14) with our experimental data, we shall rewrite it in the following form:

$$j_s = \frac{A}{\sqrt{t+t0}} - j_b , \tag{15}$$

where

$$A = C_H S \sqrt{\frac{D}{\pi}} . \tag{16}$$

We shall find parameters for Cottrell's equation (15) from the requirements of coincidence the equation with the experimental data (Figures 4 and 5) with the aid of the method of least squares and the optimization procedure by Levenberg–Marquardt. The optimization results are represented in the Table 12. Besides using the data from the Tables 11 and 12, it is possible to find diffusion coefficients for hydrogen atoms in oxide-nickel and cadmium electrodes from equation (16).

Table 12. Optimal values for parameters in Cottrell's equation

Parameters	Figure 4	Figure 5
A (ml(H_2) $s^{-1/2}$)	37.985	20.141
$t0$ (h)	11.607	6.458
j_b (ml(H_2) s^{-1})	0.070	0.055
δ^a (%)	2	4
C_H (ml(H_2) cm^{-3})	4.464×10^3	2.868×10^3
D (cm^2 s^{-1})	5.769×10^{-9}	4.154×10^{-9}

δ^a – relative error of experimental data for Cottrell's equation (15) (Figures 4 and 5)

The found diffusion coefficients (Table 12) are much less than the diffusion coefficients of free atoms of hydrogen in any metals [63]. A decrease of the diffusion coefficients is possible only in the case that the hydrogen inside of the electrodes is not in its free state (α-phase) but instead in its bound state, i.e., in the β-phase. If an atom of the hydrogen is there in its bound state (in a potential hole), a probability of hydrogen atom transmigration from one potential hole into another will decrease with grow bond strength. Hence, upon the same gradient of concentration, a diffusion coefficient will decrease, as long as the bond strength will grow.

The diffusion coefficient for hydrogen atoms in the nickel matrix of an oxide-nickel electrode found by us is about 10^6 times less than the diffusion coefficient of free hydrogen atoms in nickel at the temperature 800°C [64]. Thus, to a large extent, the hydrogen atoms are connected inside the ceramic-metal matrices of porous electrodes of nickel-cadmium batteries.

Thus, the processes of hydrogen release relaxation at a thermal decomposition of sintered electrodes of nickel-cadmium batteries are determined by the following features of the hydrogen inside the ceramic-metal matrices: Those features are, first, a distribution of atomic hydrogen all around the entire volume of the ceramic-metal matrices of the electrodes and, second, the fact that the atomic hydrogen is found inside the electrodes in its bound state, i.e., in the β-phase. These facts matter much for understanding the reasons of hydrogen accumulation in very large amounts in electrodes of nickel-cadmium batteries in a case of their long-time operation [16, 28]. This problem requires further studies, both theoretical and experimental.

3.4. Investigation of Hydrogen Release Rate from Electrodes of Nickel-Cadmium Batteries at Their Thermal Decomposition

For the examination, nickel-cadmium batteries KSX-25 were chosen with sintered electrodes with capacity of 25 Ah. The electrodes' thermal decomposition was performed in the thermal chamber (Figure 3) [65].

For the examination, batteries were chosen with an operating life of 6 years as the top amount of hydrogen is accumulated in electrodes after 3 to 5 years of operation [16, 28]. Then gradually, the oxide–nickel and cadmium electrodes were thermally decomposed in the above described installation. To the cadmium electrodes, the following temperatures were applied: 480, 600, 800°C, while to the oxide–nickel ones, the temperatures were chosen 670, 840, 880°C.

The temperature values for the electrodes decomposition were chosen based on the following considerations. In our preliminary experiments, there was found that an initiation of a noticeable gas release starts from 340°C for a cadmium electrode and from 660°C for an oxide–nickel one.

A rate of gas release was estimated based on a released gas volume for five minutes. The gas release rate measuring was conducted every hour.

The experimental results in the form of diagrams for three values of temperature are represented in Figures 6 and 7.

From the diagrams on Figures 6 and 7, it is seen that the gas release process at the thermal decomposition of the electrodes takes on average 140 hours for an oxide–nickel electrode and 78 hours for a cadmium electrode. With time, the rate of the gas release decreases exponentially. For a day, from an electrode, the following amount of gas was released: for a cadmium electrode not less than 175ml/g (milliliters of gas per one gram of an electrode weight) in the first days and down to 4 ml/g in the last days; while for an oxide–nickel electrode from 250 ml/g down to 4 ml/g. The process of thermal decomposition stopped, when the daily gas release became less than 4 ml/g.

The gas analysis was conducted with the aid of the gas analyzer VOG-2M. The analysis showed that the gas released as a result of electrodes thermal decomposition is represented by the hydrogen in total.

It is interesting to note that as a result of the electrolyte decomposition (in a course of battery charging), gases released are hydrogen and oxygen. But as it was shown by the experimental investigations, in electrodes, only the hydrogen is accumulated, while the oxygen leaves a battery.

This is connected with the fact that hydrogen possesses a very high diffusional penetrability. The diffusion coefficient of hydrogen atoms in metals is many times higher than a diffusion coefficient of any other atoms, including oxygen atoms. For example, at the temperature 20°C, the hydrogen diffusion coefficient in nickel is approximately 10^{10} times higher than the diffusion coefficient of nitrogen or oxygen [31].

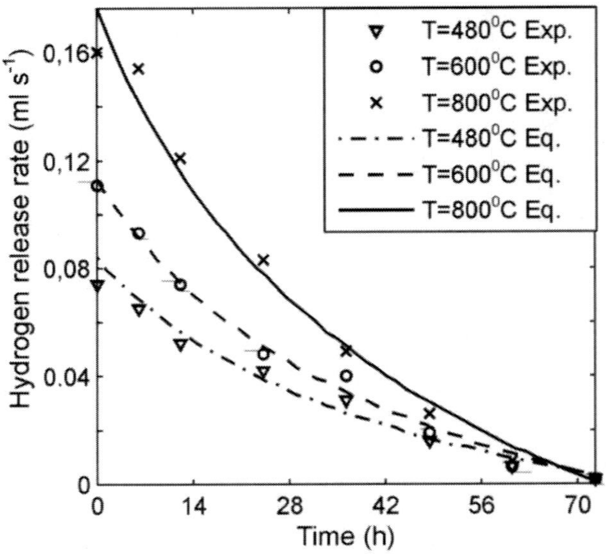

Figure 6. Rate variations of hydrogen release in a course of thermal decomposition of cadmium electrodes of battery KSX-25 at heating-up different temperatures: Exp – experimental data, Eq – Cottrell's equation (22) with parameters from Table 13.

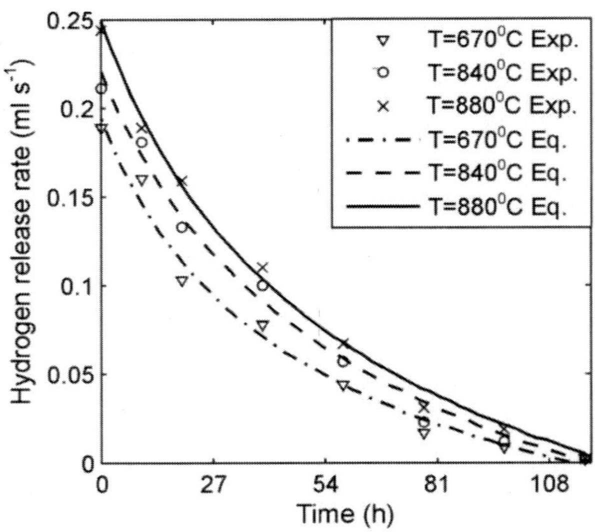

Figure 7. Rate variations of hydrogen release in a course of thermal decomposition of oxide–nickel electrodes of battery KSX-25 at heating-up different temperatures: Exp – experimental data, Eq – Cottrell's equation (22) with parameters from Table 14.

In proportion to decomposition temperature growth, a gas release rate grows (Figures 6 and 7). That is, the higher the temperature, the higher the gas release rate at any moment of time of the electrodes decomposition and hence, more hydrogen can be extracted from the electrodes. To this conclusion, one can also come by way of calculating the areas under every diagram on Figures 6 and 7, which are equal to the volumes (V) of gas released as a result of the electrodes thermal decomposition (Tables 13 and 14). But at the temperature values higher than 900^{0}C, as a matter of fact, the volume of released hydrogen does not grow any more.

This result can be explained with the supposition that hydrogen atoms in electrodes are contained in a potential hole, i.e., in their bound state. Then, according to Boltzmann distribution, the higher the temperature is, the more hydrogen molecules are able to leave the potential hole. But at temperatures higher than 900°C, all accumulated hydrogen has already left the electrodes. This fact is very important for understanding of form in which hydrogen is contained in electrodes of nickel-cadmium batteries.

The gas release at electrodes thermal decomposition is conditioned by diffusional processes; that is why it should be described by the Cottrell's equation [62]:

$$j = \frac{C_H S D}{\sqrt{\pi D t}}, \tag{17}$$

$$\gamma = \sqrt{\pi D t}, \tag{18}$$

where j – hydrogen atoms flow from electrode, C_H - hydrogen volume concentration in electrode, S – electrode surface area, D - effective diffusion factor of hydrogen atoms in ceramic-metal matrix of electrode, γ - diffusion boundary layer thickness in ceramic-metal matrix of electrode.

In equation (17), at $t \to 0$, the hydrogen atoms flow from electrode tends to infinity, which is physically meaningless. In our experiments, we started measuring the hydrogen flow from the thermal chamber, when a temperature inside of the thermal chamber reached the necessary value.

Hence by the beginning of the experiment, the diffusional boundary layer will have the thickness:

$$\gamma_0 = \sqrt{\pi D t_0} \ . \tag{19}$$

So, if to measure the hydrogen atoms flow from an electrode from the moment t_0, the equations (17,18) are necessary to be written in the form:

$$j = \frac{C_H S D}{\sqrt{\pi D(t + t_0)}}, \tag{20}$$

$$\delta = \sqrt{\pi D(t + t_0)} \ . \tag{21}$$

It is important to also take into account a possible reverse flow of hydrogen from the thermal chamber into the electrode. In the thermal chamber, the constant temperature and constant pressure [16, 28] are kept; hence the reverse flow j_b must also be constant. With due consideration of the hydrogen reverse flow, the equation (20) will have the form:

$$j = \frac{A}{\sqrt{t + t0}} - j_b, \tag{22}$$

where

$$A = C_H S \sqrt{\frac{D}{\pi}} \ . \tag{23}$$

With use of experimental data (Figures 6 and 7), let's find parameters for Cottrell's equation (22) applying the least-squares procedure and Levenberg–Marquardt's optimization procedure [66]. The results are presented in Tables 13 and 14, while the theoretical curves based Cottrell's equation (22), with parameters from Tables 13 and 14 are represented on Figures 6 and 7.

Table 13. Optimal values for cadmium electrode parameters in Cottrell's equation

Parameters	T=480⁰C	T=600⁰C	T=800⁰C
A (ml(H_2) $s^{-1/2}$)	57.04	74.977	122.51
$t0$ (h)	29.72	28.48	29.217
j_b (ml(H_2) s^{-1})	0.091	0.122	0.202
δ (%)	8	5	6
V(L)	8.181	10.649	16.05
D (cm^2 s^{-1})	4.217×10^{-8}	7.287×10^{-8}	1.946×10^{-7}

Table 14. Optimal values for oxide–nickel electrode parameters in Cottrell's equation

Parameters	T= 670⁰C	T= 840⁰C	T=880⁰C
A (ml(H_2) $s^{-1/2}$)	113.73	164	172.776
$t0$ (h)	29.16	39.42	37.03
j_b (ml(H_2) s^{-1})	0.157	0.216	0.225
δ^a (%)	5	5	6
V(L)	25.24	31.78	36.60
D (cm^2 s^{-1})	5.144×10^{-8}	1.069×10^{-7}	1.187×10^{-7}

δ^a – relative error of experimental data for Cottrell's equation (22) (Figures 6, 7).

From equation (23) with the use of data from Table 15, it is possible to find effective diffusion factors of hydrogen atoms in oxide–nickel and cadmium electrodes (Tables 13 and 14).

The values of the diffusion factors (Tables 13 and 14) represent an average effective estimation for the diffusion factors of hydrogen atoms in the electrodes during the entire process of the electrodes' thermal decomposition.

Diffusion coefficients of hydrogen atoms in solid bodies at different temperatures are well-characterized by Arrhenius equation $D=D_0\cdot\exp(-E/RT)$ [63]. Using this equation and the experimental data (Figures 6 and 7), it is possible to find the hydrogen atoms' activation energy E for the diffusion process. On Figure 8, there is represented the dependence diagram ln(D) vs. (1/T).

Table 15. Parameters of oxide–nickel and cadmium electrodes of a battery KSX-25

Type of electrode	Ni	Cd
Period of operation (years)	6	6
Electrode size (cm)	7.3×13.6×0.081	7.1×13.6×0.065
Top amount of hydrogen released (L)[a]	36.60	16.05
Hydrogen concentration in electrodes (ml(H_2) cm^{-3})	4.477×10^3	2.549×10^3
Activation energy (eV)	0.37	0.34

[a] The relative error in the data in Table 15 is 5–7%.

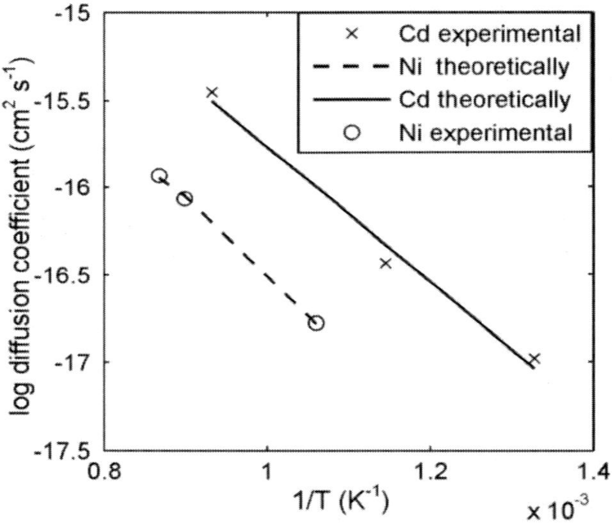

Figure 8. Dependence of hydrogen diffusion coefficient on temperature for oxide–nickel and cadmium electrodes.

From Figure 8, it is seen that the experimental data shows a good correspondence to a straight line, which confirms the applicability of the Arrhenius equation. The calculated values of the activation energy for oxide–nickel and cadmium electrodes are represented in Table 15. These values are indicative of the fact that the hydrogen is bound quite strongly inside of a ceramic-metal matrix of a porous electrode [16].

The investigations conducted, show that in the course of a long-lasting operation of nickel-cadmium batteries KSX-25 in their ceramic-metal electrodes, a great amount of hydrogen is accumulated. The hydrogen atoms are contained inside the electrodes' ceramic-metal matrix [16] in a bound state (β-phase). Notably, the connection is quite strong as the hydrogen atoms activation energy for a diffusion process is equal to approximately 0.35 eV. This can serve as an explanation of the very low diffusion coefficient of hydrogen atoms in ceramic-metal nickel matrix of electrodes (Table 14) as compared with the diffusion coefficient of non-bound hydrogen atoms (α-phase) in nickel [63].

4. Mechanism of Thermal Runaway in Alkaline Batteries

4.1. Introduction

Thermal runaway is encountered in batteries of practically any electrochemical system [1]. According to the standard point of view [1], thermal runaway in accumulators goes on with the following scenario. In the case of an accumulator long recharging under constant voltage (or at its floating service), heating-up takes place, which results in the decreasing of an accumulator's internal resistance and an increase of recharging current, which in turn magnifies heating up, etc. So the thermal runaway is a result of a positive feedback between a current and a temperature of accumulators during their charging under a constant voltage. Besides, for hermetic accumulators, there is a belief that an essential initial contribution to accumulators heating-up is made by exothermic reactions of an oxygen cycle.

In the case of the thermal runaway initiation, a battery heats up until high temperature values; an electrolyte evaporates instantly; a battery plastic body melts and sometimes it catches fire or explodes depending on the battery body construction and material.

So the thermal runaway classical mechanism supposes that the reason for the thermal runaway is an acceleration of a battery charge reaction (on expense of its heating-up) (1,2) or an acceleration of an electrolyte decomposition reaction (3,4) as the thermal runaway takes place in cases of long-time battery recharging. In Section 2.3, it was shown that a lot of experimental data exists that contradict the generally accepted thermal runaway mechanism.

In particular, according to the classical mechanism [1], the thermal runaway occurs at the expense of an energy obtained by a battery from a recharger. But the direct experimental studies show that as a result of a thermal runaway, a battery releases approximately 140 times more energy than it obtains from the recharger [17]. Therefore, the thermal runaway is connected with the development of a powerful exothermic reaction. The thermal runaway's classical mechanism does not suppose a development of some additional electrochemical reactions, except for one of battery charging and one of electrolyte decomposition (at battery recharging).

This fact can be explained from an assumption that the thermal runaway is accompanied by a powerful exothermic reaction. Moreover, this reaction must be a basis of the mechanism of thermal runaway.

As during the thermal runaway, a great mass of the hydrogen is released from the batteries for 3-5 minutes (Section 2.2 and [12, 14]), so the assumption can be made that the hydrogen is an active participant of the thermal runaway process and of the exothermic reaction accompanying this process.

Hydrogen is present inside battery electrodes in a form of metal-hydrides, i.e., in atomic form [31]. After the atomic hydrogen release from the electrodes, its recombination occurs:

$$H + H \rightarrow H_2. \tag{24}$$

The reaction (24) is a powerful exothermic reaction with heat dissipation in the amount of 436 kJ/mole (hydrogen) [66]. This heat build-up is more intensive than in a case of hydrogen combustion reaction in oxygen, which makes 285.8 kJ/mole (hydrogen) [67].

$$H_2 + \frac{1}{2}O_2 \rightarrow H_2O. \tag{25}$$

It is the atomic hydrogen recombination reaction (24) which the hydrogen welding is based on; the latter is able to create a fire up to 3700°C, while the oxygen-hydrogen welding based on the reaction (25) is able to create a fire with temperature about 2500°C [68].

So the atomic hydrogen recombination reaction is the thermal runaway exothermic reaction in question [17].

In the paper [17], it was proved that the exothermic reaction of the thermal runaway (24) proceeds via electrochemical mechanism. Indeed, in a case of a thermal runaway initiation, the battery charging current rises steeply, while the voltage on battery terminals tumbles down approximately to 0.55V (Figure 1 (No. 1) and paper [17]).

Such a sharp voltage drop cannot be explained only by the decrease in internal resistance of the battery due to increase in temperature. Actually, this voltage is even less than the OCV for the reactions (1,2) and (3,4), i.e., under the voltage 0.55V on the battery terminals these reactions would not take place at all [17]. This voltage drop can only be explained by the decrease in OCV during the process of thermal runaway. It is possible only if the thermal runaway process is related to the initiation of a new potent electrochemical reaction, with a much lower difference (ΔE) in the standard potentials for the anode and cathode, as compared to reactions (1,2) and (3,4).

In this manuscript, by experiments, we study thermal runaway and electrochemical reactions on a cathode and anode in alkaline batteries.

4.2. Examination of Electrochemical Reactions on Nickel Hydride

In Section 3 and [12, 16, 28] it was shown that as a result of a long-term operation of alkaline batteries, in their electrodes, a lot of hydrogen accumulates. So sintered matrices of long-life alkaline battery electrodes represent metal hydrides with a very high content of hydrogen [16, 28]. Subsequently, in order to understand the thermal runaway mechanism, one needs to know which electrochemical reactions take place on metal hydrides in the charge process of alkaline batteries. In this section, we shall study an electrochemical behavior of nickel hydride, which forms metal-hydride matrices of positive electrodes of nickel-cadmium batteries with a long period of operation [69].

In this section, the electrochemical behavior of the nickel hydride was examined in an electrolyte of an alkaline battery. This is the solution KOH (Reagent) by density 1.20 g cm^{-3} with the addition of 20 g L^{-1} of LiOH. In our investigation, a plate of the electrolytic nickel was used by square 1 cm^2. For creation of a reproducible surface, the plate was electropolished during 30 s in a 60% solution of H_2SO_4 at an anode current density i = 5000 A m^{-2}. Then the plate was washed with bi-distilled water and dried up with an absorbent paper. For the nickel hydride formation, the plate was placed in a cell with a solution of H_2SO_4 in concentration 1N with the addition of 0.2 g L^{-1} thiourea. The nickel hydride formation process takes place at cathode current density i_k = 500 A m^{-2} for one hour. In this case, according to the data [70], in the plate, β-phase of the nickel hydride formed. Then the electrode was washed with bi-distilled water and carried over into the working cell for an examination. The working cell was placed under a nitrogen atmosphere (99.999%) to avoid electrode oxidation. The potential of a working electrode was found versus a silver chloride electrode and then it was recalculated based on the standard hydrogen electrode (SHE).

At first, polarization curves were defined for non-hydrogenated nickel in the alkaline battery electrolyte at cathode and anode polarizations (Figure 9). Hydrogen and oxygen evolution on metals adheres well to the Tafel's equation [71, 72].

$$E = A + B \cdot \log(i), \tag{26}$$

where E – potential; i – current density on working electrode; A, B – empiric constants. The inclination of the cathode polarization curve on Figure 9 is equal to 0.12 V, which corresponds to the hydrogen evolution reaction on the cathode [73]

$$H_2O + e^- \rightarrow H_{ads} + OH^- . \quad (27)$$

The inclination of the anode polarization curve on Figure 9 is equal to 0.04V, which corresponds to the oxygen evolution reaction on the anode [73].

$$2OH^- \rightarrow H_2O + O_{ads} + 2e^- . \quad (28)$$

A similar form is taken by the polarization curves also in a case that in nickel there is only chemically free, not fixed hydrogen (α-phase of nickel hydride).

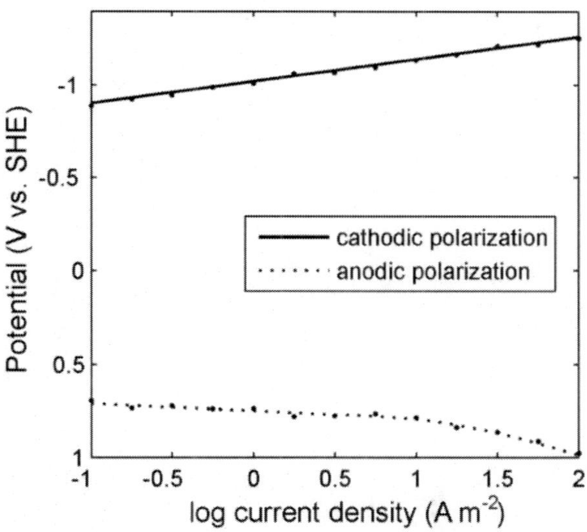

Figure 9. Polarization curves for nickel (non-hydrogenated) in alkaline battery electrolyte.

Polarization curves were then obtained on a nickel hydride in the case of presence of chemically fixed hydrogen in the metal (β-phase of nickel hydride) (Figure 10). The working samples were prepared in accordance with above described method. The polarization curves were obtained by the way of polarization multistage increase with the pitch of 0.02V in the same alkaline battery electrolyte. On the chart represented in Figure 10, it is seen that on both cathode and anode branches of the polarization curves, limiting currents are observed. This is typical for the β-phase of the nickel hydride. On the non-hydrogenated nickel as well as on the α-phase of the nickel hydride, they are absent (Figure 9).

On the cathode branch, the limiting current is observed in the area of potentials from -0.65V to -1.00V. Afterward, the chart takes the form of a straight line with the angular inclination 0.12 V, which corresponds to hydrogen evolution in accord with the reaction (27) [73]. This is to say that in this area, the hydrogen evolves from the nickel hydride exactly in the same way as on common nickel.

On the anode branch, the limiting current is observed in the area of potentials from -0.35V to 0.7V. Then, the chart has the form of an (approximately) straight line corresponding to the line of the oxygen evolution on the Figure 9.

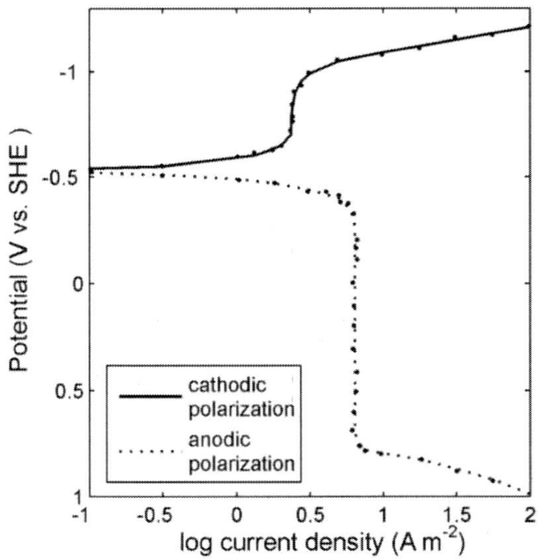

Figure 10. Polarization curves for nickel hydride in alkaline battery electrolyte.

For a better understanding of the cathode and anode processes taking place on the nickel hydride, let us consider a dependence diagram of the nickel hydride potential from time duration (Figure 11) in a case of a current absence in the cell. On parts 1 and 2 of Figure 11, some relaxation processes take place connected with hydrogen transposition from nickel hydride into the solution. Then the potential comes to equilibrium (part 3 Figure 11).

The hydrogen can arrive to the electrode surface from the α-phase of the nickel hydride, i.e., at the expense of free chemically non-fixed hydrogen atoms with aid of the diffusion in the solid. Moreover, from the β-phase of the nickel hydride, with the preliminary step of its decomposition,

$$NiH_x \rightarrow Ni + H_x . \tag{29}$$

Our additional studies at lower concentrations of alkalis (Figure 11 and others) showed that the balance potential E_r grows linearly with pH decrease and with inclination 0.06V, i.e.

$$E_r = E_o - 0.06 \cdot pH \tag{30}$$

Besides, an extrapolation of the dependence (30) onto the ordinate axis leads to a potential of the hydrogen electrode at pH=0. So E_r on a nickel hydride coincides with the potential of the hydrogen electrode [62, 71]

$$H_{ads} + OH^- \leftrightarrow H_2O + e^- . \tag{31}$$

For the further investigations, it is important to know that relaxation processes - connected with the potential coming into equilibrium condition (31) after the placing of an electrode in the solution being under studying – continue approximately for 38 minutes (Figure 11).

For working out in more detail, it is necessary to clarify some points. First, what electrochemical reactions proceed in the cases of cathode and anode polarization of the working nickel-hydride electrode? Second, under what steps do those reactions pass? Third, which of them is a rate-limiting step, i.e., a step responsible for the emergence of limiting cathode and anode currents?

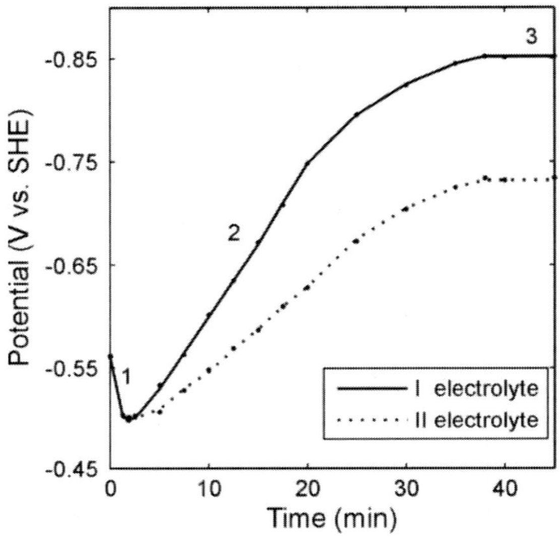

Figure 11. Dependence of nickel hydride potential from time in absence of a current in a cell: I – in a standard alkaline battery electrolyte; II - in electrolyte with alkali concentration 100 times less than in the standard battery electrolyte.

4.3. Cathode Process on Nickel Hydride

Let us study the cathode process on the nickel hydride. The atomic hydrogen being formed as a result of nickel hydride disintegration diffuses to the electrode surface and is adsorbed on it. Hence, as the electrochemical step at the case of the cathode process, only the step of the electrochemical desorption of the atomic hydrogen adsorbed on the electrode surface can be [71-73], i.e.,

$$H_2O + H_{ads} + e^- \rightarrow H_2 \uparrow + OH^- \qquad (32)$$

Now let us find out which step of the cathode process on the nickel hydride is the limiting one, i.e., one responsible for the limiting cathode current in the Figure 10. Only two steps take precedence in the cathode electrochemical reaction (32), namely the nickel hydride disintegration step (29) and the step of the atomic hydrogen diffusion to the electrode surface.

Differences of these steps can be highlighted with aid of cyclic voltammetry (CV). In Figure 12, the nickel hydride CV is represented. The cyclic voltammogram was taken in the potentials area from -0.98 V to -0.53 V, which just covers the area of the limiting cathode current (Figure 10). The scanning speed v = 0.001 V s^{-1}.

From Figure 12, it is seen that the limiting cathode current lowers with a cycle number increase. Such CV change is typical, where a previous chemical reaction is slowed down in relation to the main electrochemical reaction proceeding on the electrode [74]. In the above (considered by us) cathode process, this previous chemical reaction can only be the reaction of the nickel hydride disintegration (29).

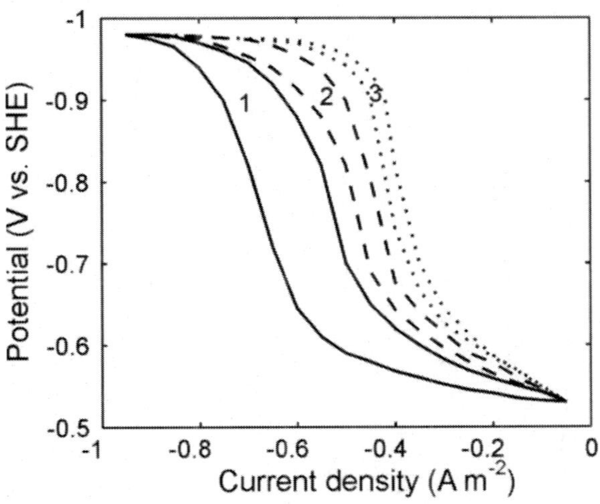

Figure 12. Cyclic voltammograms for nickel hydride in alkaline battery electrolyte obtained in potentials area from -0.98 V to -0.53 V; scanning speed is 0.001V s^{-1}; 1,2,3 – cycle numbers.

It is also possible to give other pro arguments for this statement. For kinetics clarifying of the previous limiting step, the cathode chronocoulometric diagram of the nickel hydride, i.e., the dependence between the carried over cathode charge q^- and time (Figure 13) was considered. For this purpose, in every CV cycle, a coulometer was used to find out the negative charge q^- consumed for the electrochemical desorption of the hydrogen (32). This diagram was obtained as a result of cycling in the potentials area from -1 V to -0.53 V (scanning speed 0.001 V s^{-1}).

From Figure 13, it is seen that our experimental points are well approximated with a straight line.

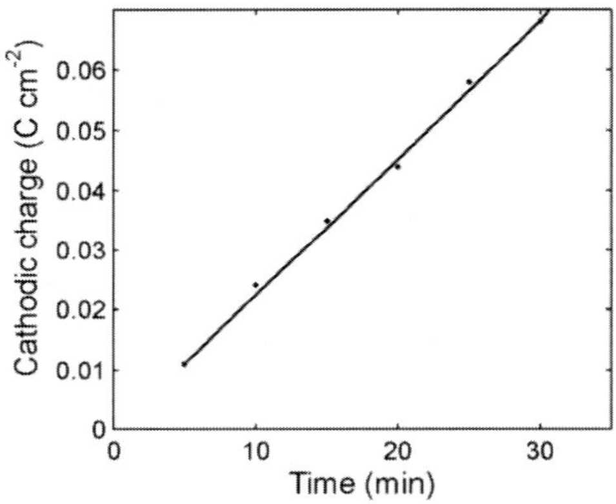

Figure 13. Cathode chronocoulometric diagram for nickel hydride in alkaline battery electrolyte obtained in cycling in the potential area from -1 V to -0.53 V.

If, to suppose that the limiting step is one of atomic hydrogen diffusion, from the Cottrell's equation [62] for the dependence of the limiting cathode current from time, we shall obtain the equation:

$$i_k = nFC_H \sqrt{\frac{D}{\pi t}} . \quad (33)$$

Then, integrating the equation on time, we shall obtain:

$$q^- = 2nFC_H \sqrt{\frac{Dt}{\pi}} . \quad (34)$$

Thus in this case, the dependence $q^{-}(t)$ must be linearized in the coordinates q^{-} vs. $\sqrt{t}$, which does not correspond to the experiment (Figure 13).

If, to suppose that the limiting step is the one of nickel hydride disintegration (29), then according to studies [75], this is a chemical reaction of the first order. For chemical reactions of the first order, a linear dependence is inherent in the coordinates q^{-} vs. t extrapolated into zero of the coordinate system, [74], which indeed is observed in the experiment (Figure 13). Thus, the limiting step of the cathode process on the nickel hydride is the step of nickel hydride disintegration (29).

4.4. Anode Process on Nickel Hydride

Now let us investigate the anode process on the nickel hydride. For clarifying the nature of the proceeding electrochemical reaction, the dependence of the electrode potential from time at the anode current density $i_a = 0.5$ A m^{-2} (Figure 14) was taken. In all the below described experiments, preliminarily, the electrode was exposed to the electrolyte action for 40 minutes with a purpose of excluding the relaxation processes (parts 1 and 2, Figure 11).

From Figure 14, it is seen that during the first 8 to 10 minutes, the potential rises steeply by 0.24V; then in about 220 subsequent minutes it continues its growth but very slowly; then again it rises steeply up to the oxygen evolution potential value.

Thus, unequivocally, the nickel hydride chronopotentiogram shows that in the area of the potential value -0.6 V, some electrochemical reaction takes place. Let us find the parameters of this reaction. Let's make use of the equation built for potential change on the chronopotentiogram from the paper [74]:

$$E = -\frac{RT}{\beta F}\ln\left(\frac{nFC_H K_H}{i}\right) - \frac{RT}{\beta F}\ln\left(1-\sqrt{\left(\frac{t}{t_c}\right)}\right). \qquad (35)$$

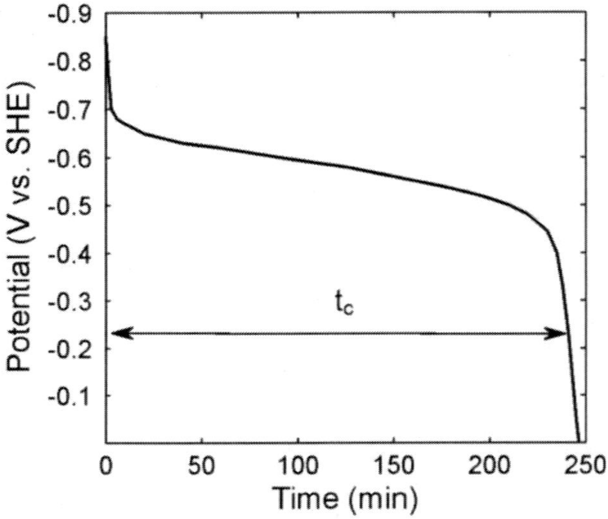

Figure 14. Chronopotentiogram for nickel hydride in alkaline battery electrolyte under anode current density $i_a = 0.5$ A/m^2, with preliminary electrode exposition in electrolyte during 40 minutes.

From this equation, under $t = 0$, we'll obtain the dependence of the electrode potential from the anode current value in the form of the Tafel's equation:

$$E = -\frac{2.3RT}{\beta F}\log\left(nFC_H K_H\right) + \frac{2.3RT}{\beta F}\log(i)\ . \qquad (36)$$

The dependence E vs. $\log(i)$ for the equation (36) was obtained experimentally. For this purpose, a number of chronopotentiograms were taken at different values of the anode current density i. The data obtained were represented on a chart in coordinate system E vs. $\log(1-\sqrt{t/t_c})$ according to equation (35). An extrapolation of these straight lines onto the ordinate axis (t=0) gives potentials E corresponding to initial hydrogen concentrations in nickel hydride at different i values. The results are represented in Figure 15.

For the inclination angle of the polarization straight line (Figure 15), the classical Tafel's value 0.12±0.005 V is obtained. From this and from equation (36), for the anode coefficient β, we obtain the value $\beta \approx 0.5$. Both obtained values correspond to electrochemical step of hydrogen ionization [71-74].

$$H_{ads} \rightarrow H^{+} + e^{-}. \tag{37}$$

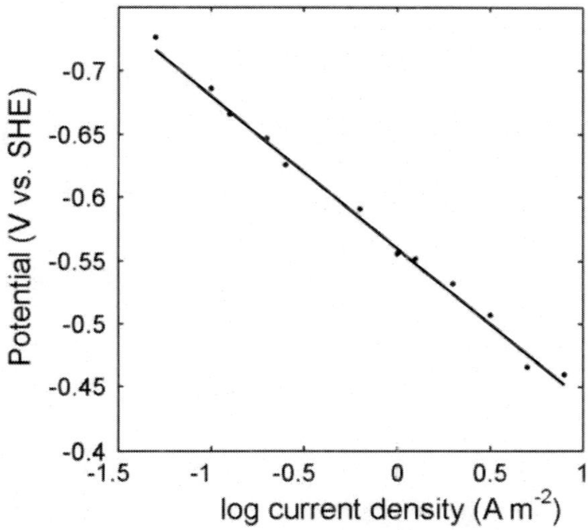

Figure 15. Anode polarization dependence for nickel hydride in alkaline battery electrolyte.

As in an alkaline environment, protons immediately interact with ions OH^-, the reaction (37) is necessary to be written down in the form:

$$H_{ads} + OH^{-} \rightarrow H_2O + e^{-}, \tag{38}$$

which does not change the sense of the proceeding electrochemical reaction.

Now let us give a consideration to the following question: what step of the anode process on the nickel hydride is the limiting step? Only two steps take precedence of the anode electrochemical reaction (38), namely the step of the nickel hydride disintegration (29) and the step of the atomic hydrogen diffusion to the electrode surface.

These step differences can be found with aid of the cyclic voltammetry (CV). For this purpose, the cyclic voltammogram of the nickel hydride was taken in the potential area from -0.55 V to 0.05 V, which covers just a part of the area of the limiting anode current (-0.35…+0.65). The potential scanning speed is v = 0.05 V s^{-1} Figure 16.

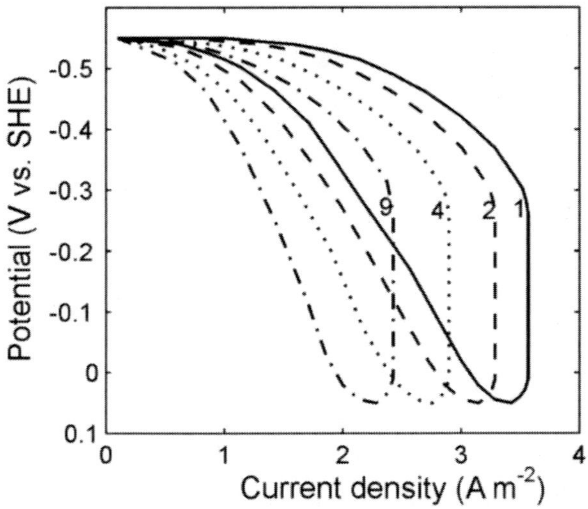

Figure 16. Cyclic voltammograms for nickel hydride in alkaline battery electrolyte obtained in the potential area from -0.55 V to 0.05 V; scanning speed 0.05 V s^{-1}; 1…9 – cycle numbers.

On the CV, there are well seen the limiting anode currents higher than the potential value -0.35 V; those currents become lower with a cycle number increase. Such CV behavior is typical, where a previous chemical step is slowed down in relation to the main electrochemical reaction [74] In the case of the considered anode process, this can be the only chemical reaction of the nickel hydride disintegration (29).

For a further clarifying of the previous limiting step, for nickel hydride, the anode chronocoulometric diagram, i.e., the dependence of carried-over anode charge q^+ from time (Figure 17) was taken. With this purpose, in every CV cycle with aid of the coulometer, the positive charge

q^+ consumed on hydrogen ionization (38) was found. This diagram was obtained at the cycling process in the potential area from -0.55 V to +0.5 V. The scanning speed was 0.05 V s^{-1}. From Figure 17, it is seen that the experimental points are well approximated by a straight line.

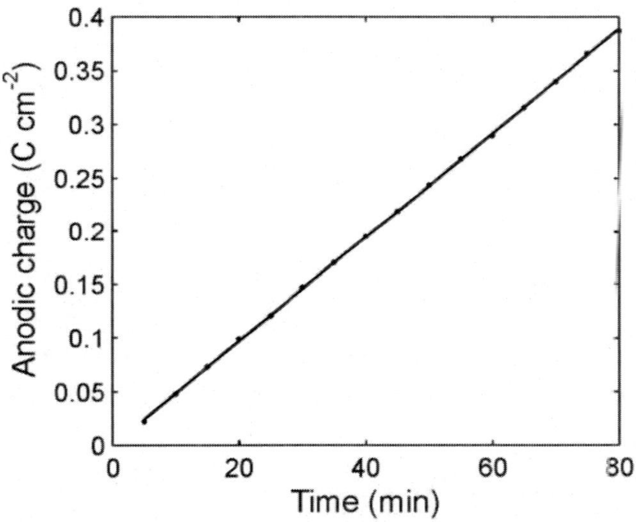

Figure 17. Anode chronocoulometric diagram for nickel hydride in alkaline battery electrolyte obtained in cycling in potential area from -0.55 V to +0.5 V.

So, if to suppose that the limiting step is one of the diffusion of the atomic hydrogen, then dependence $q^+(t)$ must be linearized in the coordinates system q^+ vs. $\sqrt{t}$ according to the equation (34), which does not correspond to the experimental data (Figure 17).

According to the study [75], the chemical reaction of the nickel hydride disintegration (29) is a reaction of the first order. For chemical reactions of the first order, the linear dependence q^+ on time is typically extrapolated onto zero of the coordinates system [74]. Such dependence is exactly observed in the experiment (Figure 17). Thus, also for the anode process on the nickel hydride, the limiting step is one of the nickel hydride disintegration (29). A speed of the nickel hydride disintegration reaction calculated on a curve inclination (Figure 17) is equal to:

$$dq^{+}/dt = 0.81 \text{ A m}^{-2}. \tag{39}$$

The equilibrium constant K for the nickel hydride disintegration (29) can be found from a number of chronopotentiograms similar to Figure 14 at different current densities on the formula [72]:

$$\frac{1}{K} = \frac{\lim\limits_{i\to 0}(i\sqrt{t_c})}{\lim\limits_{i\to\infty}(i\sqrt{t_c})} = \frac{1}{10.5 \pm 1.2}. \tag{40}$$

i.e., the reaction balance (29) is shifted to the right. Note that, as the additional analysis shows, after obtaining the chronopotentiograms, the electrolyte does not contain any nickel. Hence, in these experiments, only hydrogen is ionized, while the nickel stays in the crystal lattice.

4.5. Discussion

First of all, we would like to draw attention to a feature of electrochemical reactions (32, 38). These electrochemical reactions do not depend on a specific metal. This is why they will occur on any metals able to form hydrides. Of course, speeds of these reactions on various metals will be different as they depend on the disintegration rate of specific hydrides (29) and on the speed of hydrogen atoms diffusion in these metals.

In electrodes of long-life nickel-cadmium batteries, a lot of hydrogen is accumulated [12-17]. For example, in the batteries KSX-25 with an operation period of more than 5 years, the hydrogen is contained in amount of approximately 800 liters [12, 28]. So metal sintered matrices of electrodes of the nickel-cadmium batteries will be built of metal hydrides [16]. A thermal runaway occurs in batteries in the case of their long-time recharge [1]. Therefore, at the moment of the recharging of batteries, the following electrochemical reactions will proceed: main reactions of electrodes charging (1, 2); reactions of electrolyte decomposition (3, 4) and reactions taking place on metal hydrides (32, 38).

Let's give consideration to electrochemical reactions (32, 38) in more detail. During the charging of the nickel-cadmium batteries, the electrochemical reaction (32) will proceed on the cadmium electrode, while the reaction (38) on the oxide-nickel one will proceed. An overall reaction for them will be the following electrochemical reaction:

$$H_{ads\mathrm{Cd}} + H_{ads\mathrm{Ni}} \rightarrow H_2 \uparrow \qquad (41)$$

i.e., the electrochemical reaction of the atomic hydrogen recombination (24).

Based on the electrochemical reaction (41), it is possible to explain the experimental facts highlighted in the beginning of the paper which are impossible to be understood in the frame of the generally accepted thermal runaway mechanism [1].

Let's consider changing the main parameters of the nickel-cadmium battery (Figure 1 No. 1) in the course of the thermal runaway from the point of view of the electrochemical reaction process (41). To the moment of thermal runaway initiation in the battery KSX-25, the recharging current was very small, about 0.38A (Figure 1 No. 1). Dominantly, it was consumed on electrolyte decomposition, i.e., on electrochemical reactions (3, 4) and in a much lower percentage on the battery charge (as it was in the mode of recharging), i.e., on electrochemical reactions (1, 2). At common temperatures, a speed of electrochemical reactions (32, 38) is not high ($8.1 \cdot 10^{-5}$ A cm^{-2}) (39). Providing that electrodes square is equal to 7.3 × 13.6 cm and in the battery KSX-25, 14 cadmium and 15 oxide-nickel electrodes are situated, for reactions (32,38) maximal current, we obtain the value 0.22 A. This current is defined by hydrides disintegration speed (29) and does not depend on electrodes porosity. It is worth mentioning that with new batteries KSX-25 in similar charge modes (under constant voltage), typically, recharging current is equal to 0.15-0.18A. In new alkaline batteries, the hydrogen in electrodes is totally absent [15, 16, 28] and therefore the reactions (32, 38) are absent under the batteries' charging. In the batteries KSX-25 featured with a long operation time and thus inclined to a thermal runaway [15], typically, the recharging current is equal to 0.37-0.40A. The conducted analysis shows that the reason of the recharging current value growth is connected with electrochemical reactions occurrence (32, 38).

In the paper [17], it was shown that by taking only electrochemical reactions (1-4) into consideration, it is impossible to explain changes of charging current and voltage on battery terminals under thermal runaway (Figure 1 No. 1). Though with taking the electrochemical reactions (32, 38) into account, such parameters' change of nickel-cadmium battery becomes evident.

In proportion to nickel-cadmium batteries' number of years in operation, in their separators, dendrites are accumulated that cut a distance between electrodes exactly in the spots where they grow. This leads to the situation that at the moment of batteries charging under constant voltage, in these spots, the charging current density will be many times higher than in any other places of the same electrodes. Indeed, a local density of a charging current on electrodes in spots of dendrites growth will be:

$$i_{loc} = \frac{U_0}{\rho\ell}, \tag{42}$$

where: U_0 – voltage on battery terminals, ρ - electrolyte specific resistance, ℓ - distance between a dendrite and counterelectrode. This charging current local density i_{loc} will greatly exceed the charging current average density i_{av} on electrodes as:

$$\frac{i_{loc}}{i_{av}} = \frac{L}{\ell}, \tag{43}$$

where L – distance between electrodes. The distance ℓ between a dendrite and a counter-electrode can be tens and hundreds times less than the distance L between the electrodes. Hence, also the charging current local density in spots of dendrites growth can be tens and hundreds times more than the charging current's average density on the electrodes. In its turn, it will lead to a strong local overheating of those spots on both battery electrodes.

As shown in the paper [12], under electrodes temperature growth, metal hydrides disintegration speeds (29) will rise steeply. For a cadmium electrode at temperatures higher than 340°C, a rapid disintegration of cadmium hydrides starts, while for oxide-nickel electrodes, at temperatures higher than 660°C [12] and even at temperatures half as much, the hydrides disintegration speed is quite noticeable. The hydrides disintegration (29) (according to the above studies) is a limiting step for electrochemical reactions (32, 38). So the speeds of the electrochemical reactions (32, 38) can increase hundreds of times in proportion to the metal hydrides disintegration growth speed. The above mentioned high temperatures are not impossible for nickel-cadmium batteries. For example, the temperature evaluation in the spots of the thermal runaway as judged by thermal traces on the electrodes (Figure 2.) showed that during the thermal runaway in these spots, the temperature was much higher than 700°C. So the role of the electrochemical reactions (32, 38) can become determinative (during batteries charging) at high temperatures that can be created in local spots on electrodes at the expense of dendrites intergrowth.

The electrochemical reactions taking place on metal hydrides (41) is the exothermic reaction proceeding with very high heat production (436 kJ/mole) (hydrogen) [66]. This is why its initiation in a spot of a dendrite intergrowth will lead to even higher electrodes local heating-up in those spots; and in turn this will lead to even more mass disintegration of hydrides on both electrodes and thus to reaction speeds increase (32, 38). The exothermic reaction (41) requires new hydrides for its continuation. In a spot of the reaction initiation (41), they will be exhausted; and reaction will shift radially from this point, burning a separator in a form of regular circles, which indeed is observed in the experiments (Figure 2 and paper [17]).

An interruption of the electrochemical reactions of the thermal runaway (32, 38), in our opinion, is connected with electrolyte evaporation between electrodes (as a result of high temperature) as well as with a gas lock formation between those electrodes (from the steam evolved and the hydrogen). This gas lock will physically interrupt any electrochemical reactions (32, 38). As the formation of the gas lock sufficient for electrochemical reactions interruption, (32, 38) is dominantly a random process, hence also the radii of separator burning-out will be various, in accord with the experimental results [17].

As for the thermal runaway process, the electrochemical reactions (32, 38) become determinative, while other reactions (1-4) contributions are desperately insignificant, so on battery terminals, an electric potential difference will be established corresponding to the proceeding of these reactions. According to the above studies, during the electrochemical reaction (38), the potential of the positive electrode will be equal in average to -0.6V vs. SHE according to the chronopotentiogram (Figure 14). If taking a similar chronopotentiogram for the cathode current direction, during first 5-8 minutes the potential will tumble sharply about 0.2V starting from the balance potential -0.852V (Figure 11), then slowly decrease along the cathode branch (Figure 9). So, during the electrochemical reaction (32), the potential of the negative electrode will be, on average, equal to -1.08 V vs. SHE (Figures 9 and 10). So this electric potential difference between anode and cathode for the reactions (32, 38) coincides with the experimental voltage value on battery terminals during a thermal runaway (Figure 1 No. 1) taking into account the additional voltage drop on ohmic resistance of electrolyte and of battery electrodes. Note that at the voltage value 0.5-0.6V on the battery terminals during a thermal runaway (Figure 1 No. 1), reactions (1-4) will not proceed at all [17].

Based on the electrochemical reaction (41), it is also possible to explain other earlier difficult to understand results connected with the thermal runaway. In the paper [17], by means of the calorimetric measurement, it was shown that during the thermal runaway (Figure 1 No. 1) from the battery, the energy is produced in the amount of 5207 kJ. Notably, during the thermal runaway, the amount of the energy produced exceeds the energy amount received by the battery from a recharger 140 times [17]. This experimental fact cannot be explained by the generally accepted thermal runaway mechanism [1], which takes into account no other electrochemical reactions except of the reactions (1-4). Based on the exothermic reaction (41), this fact has a quite natural explanation. The electrochemical reaction (41) explains the entire complex of the experimental studies connected with the thermal runaway.

5. Analytical Model of Thermal Runaway in Alkaline Batteries

5.1. Introduction

It is a generally accepted concept [1] that the cause of thermal runaway is a battery self-heating in the course of its charging at constant voltage or in the course of its work in floating mode. Heating of a battery results in a decreasing of its internal resistance and therefore its charging current grows, which again increases the heat, etc. In this case, the battery temperature goes up to high values; also inflammation and explosion are possible. Any thermal runaway always results in the breakdown of a system containing this battery.

However, until now there is no direct experimental proof of this mechanism. According to the traditional point of view [1], the thermal runaway takes place due to acceleration of batteries' charging reactions (1, 2) or to acceleration of electrolyte decomposition reactions (3, 4), as the thermal runaway takes place in cases of a long-time battery recharging [1].

In Section 2.3, it was shown that at present, a lot of experimental facts exist contradicting the thermal runaway generally accepted mechanism. For example, at a thermal runaway, energy is produced in an amount 140 times more than what the battery obtains from its recharger. Besides, during a thermal runaway a lot of gas releases, 95% of which makes hydrogen. Also a lot of other experimental facts [12-17] exist, which is impossible to explain based on the electrochemical reactions (1-4).

In Section 4 by experiments, it was shown that the thermal runaway is connected with an initiation of a powerful exothermic reaction of atomic hydrogen recombination (41). So the thermal runaway proceeds according to the following scenario. During long-lasting batteries' operation, into their electrodes, a lot of hydrogen is accumulated in the form of metal-hydrides (i.e., in the atomic form [31]), while on cadmium electrodes, dendrites are accumulated. In spots of dendrites penetration through a separator, the distance between the electrodes is reduced considerably. Hence, in those spots, the current density of the batteries' charge will be much higher than in neighboring areas of the electrodes. A rate-limiting step for the electrochemical reactions (32, 38) is the metal-hydrides decomposition step (29). The metal-hydrides decomposition rate grows sharply with electrodes temperature growth. Thus, in the dendrite location spot, an intensity of the reactions (32, 38) will grow. As the thermal runaway reaction (41) is an exothermic reaction, it will heat up the electrode even more. This will result in a more intensive metal-hydrides decomposition, etc. So the reactions' (32, 38) intensity will rise steeply. This is exactly what we know as a thermal runaway process in alkaline batteries [69].

With time, the thermal runaway process stops because of a gas lock formation between electrodes, which physically interrupts the electrochemical reactions (32, 38). The gas lock is formed because of a release of a lot of hydrogen in the spot of the thermal runaway (41) and electrolyte evaporation as a result of the high temperature.

It should be noted that currently there is no reliable model of thermal runaway in batteries.

The most fundamental models of batteries are based on the electrochemical modelling method. Under this method, battery models are built based on fundamental physical and electrochemical laws [76–80]. Although these models describe all the processes in batteries with the highest accuracy, their use in technical systems is not always possible. Hausmann [81] highlighted three of the shortcomings of these models. First, they contain a lot of local parameters, which are very difficult or impossible to measure experimentally. For example, it is impossible to measure directly the various kinetic parameters inside a porous electrode, particularly the parameters of the Butler-Volmer function. Second, these models have a very sophisticated system of calibration. That is why if, at projecting of technical systems containing batteries, it may become necessary to change batteries' requirements, in this case, all experimental studies must be redone from the very beginning. Third, these models require considerable computational potency, which is unacceptable for the on-board computers of electrical vehicles and airplanes. All these factors restrict the applicability of these models.

Further, many companies require the development of such battery models, the parameters of which could be measured without disassembling a battery. It is principally impossible to build such models under the electrochemical method of modelling. These models can be developed only by using either the statistical method [82-88] or the non-linear structural method [89]. Under the non-linear structural method, the object or process under examination is considered as a system consisting of its subsystems or elements. The separation of the studied process into elements takes place by studying the process and differentiating its various stages. To every stage of the process, one or a few structural elements are assigned. The structural elements are connected by linear links based on Kirchhoff's laws. Unfortunately, this modeling method is used extremely seldomly [90, 91]. That is why in practice, statistical models of batteries are used very often [81-88]. These models are used where it is impossible or undesirable to use electrochemical models [81, 92] or where the processes being modelled are studied poorly (for example, thermal runaway).

In this section based on the found mechanism of the thermal runaway (Section 4 and [69]), the analytical model of this process is developed [93].

5.2. Analytical Model of Thermal Runaway

From the moment of thermal runaway initiation, the reactions (32, 38) rate grows as an avalanche. Indeed, the more current of the thermal runaway reactions (32, 38), the more temperature in the spot of the thermal runaway (by virtue of the fact that the thermal runaway reaction (41) is exothermic) and so the more hydrides decomposition rate and the higher rate of hydrogen release. The metal-hydrides decomposition reaction is the rate-limiting step for a generalized reaction of the thermal runaway (41) [69]. As (according to the study (Section 4 and [69])) the hydrides decomposition reaction is a reaction of the first order, the following equation must be correct for the thermal runaway reaction rate.

$$\frac{dI_{tr}}{dt} = kI_{tr}, \tag{44}$$

where I_{tr} is the current of thermal runaway and k is the thermal runaway reaction rate constant. From the equation (44), it follows that from the thermal runaway initiation moment, the current will grow exponentially

$$I_{tr} = A\exp(kt). \tag{45}$$

Now let us take into consideration a blocking of the thermal runaway reaction by a gas lock being created in a burned-down separator [17] in the spot of the thermal runaway due to the hydrogen release and the electrolyte evaporation because of the very high temperature of the thermal runaway. The thermal runaway reaction rate constant k must depend on a volume V of the gas lock as it blocks the place of the thermal runaway and makes difficult or completely interrupts the thermal runaway reactions (32, 38) on separate spots. Notably, the greater the volume of the gas lock is (i.e., the more spacious thermal runaway area is blocked), the less the reaction rate constant k must be. When the gas lock volume reaches some value V_0, the current rise rate dI_{tr}/dt must become equal to zero. Thereafter, the thermal runaway current must diminish – i.e., $dI_{tr}/dt < 0$. Hence, the thermal runaway reaction rate constant must take the form:

$$k = k_0(V_0 - V(t)), \tag{46}$$

where *V(t)* is the gas lock volume. In its turn, the gas lock volume must be in proportion with a volume of the gas released, which is proportional to an electricity amount spent on the thermal runaway reaction, i.e.,

$$V(t) = A\int_{t0}^{t} I_{tr}(t)dt\,, \tag{47}$$

where *t0* is the time moment of thermal runaway initiation and *A* is proportionality constant.

This way, with a due account of the ratios (46, 47) from the equation (44), we obtain the equation describing thermal runaway current change in view of the blocking gas lock formation:

$$\frac{dI_{tr}}{dt} = k_0(V_0 - A\int_{t0}^{t} I_{tr}(t)dt)i_{mp}(t)\,. \tag{48}$$

This equation is a differential-integral one and for its solving, it is convenient to convert it into a purely differential equation. With this purpose in mind, in the right part of Equation (48), we take the integral separately and differentiate the obtained ratio. We'll obtain the differential equation:

$$I_{tr}\frac{d^2 I_{tr}}{dt^2} - \left(\frac{dI_{tr}}{dt}\right)^2 = BI_{tr}^{\ 3}, \qquad B = k_0 A\,. \tag{49}$$

In the peak point of the thermal runaway current $t = t_k$, the following conditions must be correct:

$$I_{tr}(t)\big|_{t=t_k} = I_{\max},$$
$$\left.\frac{dI_{tr}}{dt}\right|_{t=t_k} = 0\,. \tag{50}$$

Let us solve Equation (49) at the boundary conditions (50). For the thermal runaway current change, we obtain the following expression:

$$I_{tr}(t) = I_{max}\left(1 - th^2\left(\sqrt{\frac{I_{max}B}{2}}\left(t - t_k\right)\right)\right). \tag{51}$$

Qualitatively, the theoretical curve of the thermal runaway current change (51) corresponds quite well to the experimental data obtained in the works [17, 19] and (Figure 1).

As highlighted in the papers [12, 17, 19] and (Figure 1 No. 2), often at a thermal runaway, not one but several current bursts are observed, which correspond to the rising and dying out of various focal points of a thermal runaway during batteries recharging. Most often, there can be one or two current bursts, but it is possible to observe up to five of them. In this case, the thermal runaway total current will look like:

$$I_{tr}(t) = \sum_{j=1}^{n} I_{j,max}\left(1 - th^2\left(\sqrt{\frac{I_{j,max}B_j}{2}}\left(t - t_{j,k}\right)\right)\right), \tag{52}$$

where n is the thermal runaway bursts number at batteries recharging.

The thermal runaway in batteries occurs in a case of their recharging at constant voltage [1]. In this case, the voltage on the batteries' terminals during the thermal runaway will look as follows:

$$U(t) = U_0 - R \cdot I_{tr}(t), \tag{53}$$

where U_0 is constant voltage of battery charge and R is resistance of supply lines and an instrument shunt [17]. Finally with due account to the Equation (52) for voltage change on battery terminals, we shall obtain the expression:

$$U(t)=U_0-R\cdot\sum_{j=1}^{n}I_{j,\max}\left(1-th^2\left(\sqrt{\frac{I_{j,\max}B_j}{2}}\left(t-t_{j,k}\right)\right)\right). \tag{54}$$

5.3. Experimental

In order to compare the obtained models (52, 54) with the experimental data, the nickel-cadmium batteries KSX-25 (by capacity 25 Ah) with long operating life (more than 6 years) were used. Notably, batteries' cycling was performed in severe conditions. This allowed for increasing the probability of thermal runaway emergence [15]. Batteries' charging was performed for 10 hours at high constant voltage 1.87V. Then discharging was performed by the current 10 A down to the voltage value 1 V. At the same time, ten parallel-connected batteries were cycled – for obtaining a greater amount of statistical data. Between the batteries, wooden spacer plates (two centimeters thick) were inserted. This allowed to exclude a thermal mutual influence between the batteries.

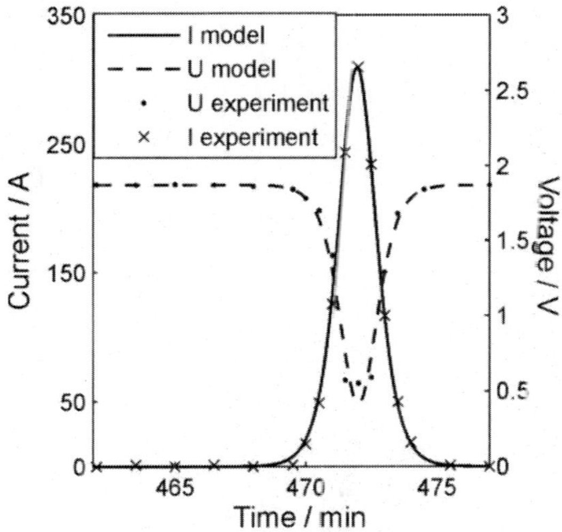

Figure 18. Comparison between analytical models for current and voltage changing and experimental data for the battery KSX-25 (No. 1) during thermal runaway: model – theoretical curves (52, 54), experiment – experimental data.

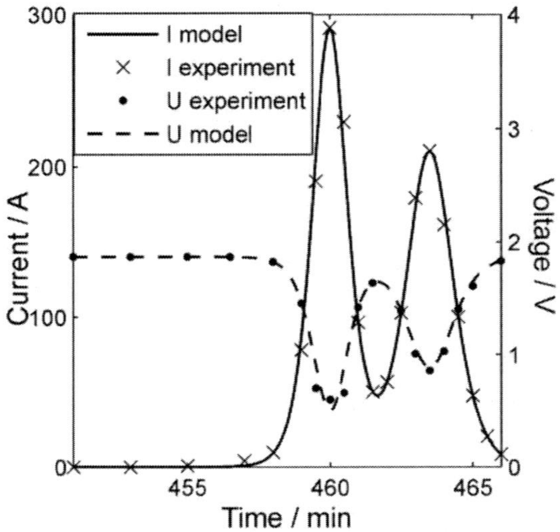

Figure 19. Comparison between analytical models for current and voltage changing and experimental data for the battery KSX-25 (No. 2) during thermal runaway: model – theoretical curves (52, 54), experiment – experimental data.

The current was measured with aid of shunts (with resistance 0.001 Ohm) fastened to the negative terminals of batteries. As a result of the cycling, there were 800 charge-discharge cycles performed. Notably, only in two cases, the thermal runaway was observed. The obtained experimental values for current and voltage changing are marked in Figures 18 and 19 by crosses and points respectively.

For Equations (52, 54) verification, the experimental data were used for changes of current and voltage on battery terminals during the thermal runaway (Figures 18 and 19).

The empiric constants (t_k, I_{max}, B, U_0, R) for these equations were found based on the requirement of the best coincidence of Equations (52, 54) with the experimental data, for this the least squares method and Levenberg–Marquardt optimization procedure were used. The optimization results are represented in the Table 16.

Thus, the obtained equations (52, 54) match well with the experimental data for current and voltage changes during a thermal runaway as the relative error of experimental data simultaneous approximation by Equations (52, 54) on the Figures 18 and 19 makes not more than 4.2%.

Table 16. Optimal values for parameters in Equations (52, 54)

Parameters	Figure 18	Figure 19	
n^a	1	1	2
t_k (min)	472	460	463.5
I_{max} (A)	310	290	210
B (A^{-1} min^{-2})	0.007	0.01	0.008
U_0 (V)	1.87	1.87	1.87
R (Ω)	0.0047	0.0047	0.0047
δ^b(%)	3.7	4.2	

n^a is a sequence number of a current burst on Figures 18 and 19.

δ^b is relative error of experimental data simultaneous approximation by the Equations (52, 54) on Figures 18 and 19.

Perhaps, the thermal runaway is the only phenomenon in batteries that is absolutely sporadic regardless of a maintaining team and being accompanied with powerful energetic and visual effects. In work [93] for the first time, there was developed the analytical model of this (from an electrochemical point of view) unique phenomenon on the basis of the recently confirmed thermal runaway mechanism in batteries with aqueous electrolyte (Section 4). Undoubtedly, this phenomenon requires further theoretical and experimental studies.

REFERENCES

[1] Guo, Y. 2009. "Thermal Runaway." In *Encyclopedia of Electrochemical Power Sources,* edited by J. Garche, 4:241-254. Amsterdam: Elsevier.

[2] Abada, S., Marlair, G., Lecocq, A., Petit, M., Sauvant-Moynot, V., and Huet, F. 2016. "Safety focused modeling of lithium-ion batteries: A review." *Journal of Power Sources* 306:178-192.

[3] Larsson, F., and Mellander, B.-E. 2014. "Abuse by external heating, overcharge and short circuiting of commercial lithium-ion battery cells." *Journal of The Electrochemical Society* 161 (10):A1611-A1617.

[4] Galushkin, N. E., Yazvinskaya, N. N., and Galushkin, D. N. 2018. "Mechanism of thermal runaway in lithium-ion cells." *Journal of The Electrochemical Society* 165 (7):A1303-A1308.

[5] Larsson, F., Anderson, J., Andersson, P., and Mellander, B.-E. 2016. "Thermal modelling of cell-to-cell fire propagation and cascading thermal runaway failure effects for lithium-ion battery cells and modules using fire walls." *Journal of The Electrochemical Society* 163 (14):A2854-A2865.

[6] Feng, X., Fang, M., He, X., Ouyang, M., Lu, L., Wang, H., and Zhang, M. 2014. "Thermal runaway features of large format prismatic lithium ion battery using extended volume accelerating rate calorimetry." *Journal of Power Sources* 255:294-301.

[7] Ren, F., Cox, T., and Wang, H. 2014. "Thermal runaway risk evaluation of Li-ion cells using a pinch–torsion test" *Journal of Power Sources* 249:156-162.

[8] Escobar-Hernandez, H. U., Gustafson, R. M., Papadaki, M. I., Sachdeva, S., and Mannan, M. S. 2016. "Thermal runaway in lithium-ion batteries: incidents, kinetics of runaway and assessment of factors affecting its initiation." *Journal of The Electrochemical Society* 163 (13):A2691-A2701.

[9] Torabi, F., and Esfahanian, V. 2011. "Study of thermal–runaway in batteries I. theoretical study and formulation." *Journal of The Electrochemical Society* 158:A850- A858.

[10] Culpin, B. 2004. "Thermal runaway in valve-regulated lead-acid cells and the effect of separator structure." *Journal of Power Sources* 133:79-86.

[11] Salkind, A. J., and Duddy, J. C. 1962. "The thermal runaway condition in nickel-cadmium cells and performance characteristics of sealed light weight cells." *Journal of The Electrochemical Society* 109:360- 364.

[12] Galushkin, D. N., Yazvinskaya, N. N., and Galushkin, N. E. 2008. "Investigation of the process of thermal runaway in nickel-cadmium accumulators." *Journal of Power Sources* 177:610-616.

[13] Yazvinskaya, N. N., Galushkin, N. E., Galushkin, D. N., and Galushkina I. A. 2016. "Hydrogen amount estimation in electrodes of nickel-cadmium batteries depending on their operating life." *International Journal of Electrochemical Science* 11:7843-7848.

[14] Galushkin, N. E., Yazvinskaya, N. N., Galushkin, D. N., and Galushkina, I. A. 2014. "Thermal runaway in sealed alkaline batteries." *International Journal of Electrochemical Science* 9:3022-3028.

[15] Galushkin, N. E., Yazvinskaya, N. N., Galushkin, D. N., and Galushkina, I. A. 2014. "Causes analysis of thermal runaway in nickel-cadmium accumulators." *Journal of The Electrochemical Society* 161:A1360-A1363.

[16] Galushkin, N. E., Yazvinskaya, N. N., Galushkin, D. N., and Galushkina, I. A. 2014. "Oxide-nickel electrodes as hydrogen storage units of high-capacity." *International Journal of Hydrogen Energy* 39:18962-18965.

[17] Galushkin, N. E., Yazvinskaya, N. N., and Galushkin, D. N. 2015. "The mechanism of thermal runaway in alkaline batteries." *Journal of The Electrochemical Society* 162:A749-A753.

[18] Galushkin, N. E., Yazvinskaya, N. N., and Galushkin, D. N. 2016. "Thermal runaway as a new high-performance method of desorption of hydrogen from hydrides." *International Journal of Hydrogen Energy* 41:14813-14819.

[19] Galushkin, N. E., Yazvinskaya, N. N., Galushkin, D. N., and Galushkina, I. A. 2016. "Analysis of thermal runaway aftereffects in nickel-cadmium batteries." *International Journal of Electrochemical Science* 11:10287-10295

[20] Vincent, C. A., and Scrosati, B. 1997. *Modern Batteries*. Oxford: Butterworth-Heinemann.

[21] Torabi, F., and Esfahanian, V. 2013. "Study of thermal–runaway in batteries II. The Main Sources of Heat Generation in Lead-Acid

Batteries." *Journal of The Electrochemical Society* 160 (2):A223-A234.

[22] Atkins, P. 1997. *Physical Chemistry*. New York: W. H. Freeman and Company.

[23] Wendt, H. 1990. *Electrochemical Hydrogen Technologies. Electrochemical Production and Combustion of Hydrogen.* Amsterdam: Elsevier.

[24] Ni, M., Leung, M., Leung, D. 2008. "Technological development of hydrogen production by solid oxide electrolyzer cell (SOEC)." *International Journal of Hydrogen Energy* 33:2337-2354.

[25] Galushkin, N. E., Yazvinskaya, N. N., Galushkin, D. N., and Galushkina, I. A. 2015. "Probability investigation of thermal runaway in nickel-cadmium batteries with sintered, pasted and pressed electrodes." *International Journal of Electrochemical Science* 10:6645-6650.

[26] Yazvinskaya, N. N., Galushkin, N. E., Galushkin, D. N., and Galushkina I. A. 2016 "Study of effect of batteries capacity on probability of thermal runaway occurrence." *International Journal of Electrochemical Science* 11:8163-8168.

[27] Yazvinskaya, N. N., Galushkin, N. E., Galushkin, D. N., and Galushkina I. A. 2016. "Probability investigation of thermal runaway in nickel-cadmium batteries with pocket electrodes." *International Journal of Electrochemical Science* 11:5850-5854.

[28] Galushkin, N. E., Yazvinskaya, N. N., and Galushkin, D.N. 2013. "Ni-Cd batteries as hydrogen storage units of high-capacity." *ECS Electrochemistry Letters* 2:A1-A2.

[29] Galushkin, N. E., Yazvinskaya, N. N., and Galushkin, D. N. 2018. "Preparation and Application of Metal Hydrides." In *Hydrides: Types, Bonds and Applications*, edited by C. D. Patrick, 193-229. New York: Nova Scince Publishers, Inc.

[30] Galushkin, N. E., Yazvinskaya, N. N., and Galushkin, D. N. 2018. "Electrochemical methods of hydrogen absorption/desorption for metal hydrides." In *Hydrogen Storage: Preparation, Applications*

and Technology, edited by Huaiyu Shao, 197-223. New York: Nova Scince Publishers, Inc.

[31] Broom, D. P. 2011. *Hydrogen storage materials the characterisation of their storage properties*. London: Springer-Verlag.

[32] Sakintuna, B., Lamari-Darkrim, F., and Hirscher, M. 2007. "Metal Hydride Materials for Solid Hydrogen Storage: A Review." *International Journal of Hydrogen Energy* 32:1121-1140.

[33] Kojima, Y., Kawai, Y., and Haga, T. 2006. "Magnesium-based nano-composite materials for hydrogen storage." *Journal of Alloys and Compounds* 424:294-298.

[34] Shang, C.X., Bououdina, M., Song, Y., and Guo. Z.X. 2004. "Mechanical alloying and electronic simulations of (MgH_2+M) systems (M=Al, Ti, Fe, Ni, Cu and Nb) for hydrogen storage." *International Journal of Hydrogen Energy* 29:73-80.

[35] Oelerich, W., Oelerich, B., Klassen, T., and Bormann, R. 2001. "Metal oxides as catalysts for improved hydrogen sorption in nanocrystalline Mg-based materials." *Journal of Alloys and Compounds* 315:237-242.

[36] Polanski, M., and Bystrzycki, J. 2009. "Comparative studies of the influence of different nanosized metal oxides on the hydrogen sorption properties of magnesium hydride." *Journal of Alloys and Compounds* 486:697-701.

[37] Borgschulte, A., Lohstroh, W., Westerwaal, R. J., Schreuders, H., Rector, J. H., Dam, B., and Griessen R. 2005. "Combinatorial method for the development of a catalyst promoting hydrogen uptake." *Journal of Alloys and Compounds* 404–406:699-705.

[38] 1986. *Hydrogen Systems*, 1:550-553. Oxford: Pergamon Press.

[39] Galushkin, N. E., Yazvinskaya, N. N., and Galushkin, D. N. 2017. "Pocket electrodes as hydrogen storage units of high-capacity." *Journal of The Electrochemical Society,* 164 (12):A2555-A2558.

[40] Byung-Joo, K., and Soo-Jin, P. 2011. "Optimization of the pore structure of nickel/graphite hybrid materials for hydrogen storage." *International Journal of Hydrogen Energy,* 38 (1):648-653

[41] Zhou, L., Meng, J., Li, P., Tao, Z., Mai L., and Chen, J. 2017. "Ultrasmall cobalt nanoparticles supported on nitrogen-doped porous carbon nanowires for hydrogen evolution from ammonia borane." *Materials Horizons* 4: 268-273.

[42] Li, X., Zhu, H., Xu, C., Mao, Z., and Wu, D. 2003. "Measuring hydrogen storage capacity of carbon nanotubes by tangent-mass method." *International Journal of Hydrogen Energy* 28:1251-1253.

[43] Pan, W., Zhang, X., Li, S., Wu, D., and Mao, Z. 2005. "Measuring hydrogen storage capacity of carbon nanotubes by high-pressure microbalance." *International Journal of Hydrogen Energy* 30: 719-722.

[44] Mu, S., Tang, H., Qian, S., Pan, M., and Yuan, R. 2006. "Hydrogen storage in carbon nanotubes modified by microwave plasma etching and Pd decoration." *Carbon* 44:762-767.

[45] Sankaran, M., Viswanathan B., and Srinivasa, M. 2008. "Boron substituted carbon nanotubes—How appropriate are they for hydrogen storage?" *International Journal of Hydrogen Energy* 33:393-403.

[46] Iyakutti, K., Kawazoe, Y., Rajarajeswari, M., and Surya, V. J. 2009. "Aluminum hydride coated single-walled carbon nanotube as a hydrogen storage medium?" *International Journal of Hydrogen Energy* 34:370-375.

[47] Weinberger, B., and Darkrim-Lamari, F. 2009. "High pressure cryo-storage of hydrogen by adsorption at 77K and up to 50MPa" *International Journal of Hydrogen Energy* 34:3058-3064.

[48] Fichtner, M. 2005. "Nanotechnological Aspects in Materials for Hydrogen Storage." *Advanced Engineering Materials* 7:443-455.

[49] Yang, R. T. 2000. "Hydrogen storage by alkali-doped carbon nanotubes–revisited." *Carbon* 38:623–641.

[50] Cabria, I., Lopez M. J., and Alonso, J. A. 2011. "Simulation of the hydrogen storage in nanoporous carbons with different pore shapes." *International Journal of Hydrogen Energy* 36:10748-10759.

[51] Gogotsi, Y., Portet, C., Osswald, S., Simmons, J. M., Yildirim, T., Laudisio G., and Fischer, J. E. 2009. "Importance of pore size in

high-pressure hydrogen storage by porous carbons." *International Journal of Hydrogen Energy* 34:6314-6319.

[52] de la Casa-Lillo, M.A., Lamari-Darkrim, F., Cazorla-Amoros, D., and Linares-Solano, A. 2002. "Hydrogen Storage in Activated Carbons and Activated Carbon Fibers." *The Journal of Physical Chemistry B* 106:10930-10934.

[53] Denisov, E. A., and Kompaniets, T. N. 2001. "Interaction of graphite with atomic hydrogen." *Technical Physics* 46: 240–244.

[54] Tazhibaeva, I. L., Klepikov, A. K., Shestakov, V. P., Romanenko, O. G., Chikhray, E. V., Kenzhin, E. A., Cherepnin, Y. S., Tikhomirov, L. N., and Zverev, V.A. 1996. "Hydrogen release of reactor irradiated RGT-graphite." *Journal of Nuclear Materials* 233-237: 1198-1201.

[55] Kanashenko, S. L., Gorodetsky, A. E., Chernikov, V. N., Markin, A. V., Zakharov, A. P., Doyle, B. L., and Wampler, W. R. 1996. "Hydrogen adsorption on and solubility in graphites." *Journal of Nuclear Materials* 233-237:1207-1212.

[56] Surya, V. J., Iyakutti, K., Venkataramanan, N., Mizuseki H., and Kawazoe, Y. 2010. "The role of Li and Ni metals in the adsorbate complex and their effect on the hydrogen storage capacity of single walled carbon nanotubes coated with metal hydrides, LiH and NiH_2." *International Journal of Hydrogen Energy* 35:2368-2376.

[57] Yürüm, Y., Taralp, A., and Veziroglu, T. N. 2009. "Storage of hydrogen in nanostructured carbon materials." *International Journal of Hydrogen Energy* 34:3784-3798.

[58] Imamura, H., Kitazawa, I., Tanabe Y., and Sakata, Y. 2007. "Hydrogen storage in carbon/Mg nanocomposites synthesized by ball milling." *International Journal of Hydrogen Energy* 32:2408-2411.

[59] Choi, M., and Ryoo, R. 2007. "Mesoporous carbons with KOH activated framework and their hydrogen adsorption." *Journal of Materials Chemistry* 17:4204-4209.

[60] Fierro, V., Zhao, W., Izquierdo, M. T., Aylon, E., and Celzard, A. 2010. "Adsorption and compression contributions to hydrogen storage in activated anthracites." *International Journal of Hydrogen Energy* 35:9038-9045.

[61] Yazvinskaya, N. N., Galushkin, N. E., Galushkin, D. N., and Galushkina I. A. 2017. “Processes of hydrogen release relaxation at thermal decomposition of electrodes of nickel-cadmium batteries.” *International Journal of Electrochemical Science* 12:2791-2797.

[62] Zoski, C. G., Ed. 2007. *Handbook of Electrochemistry*. Amsterdam: Elsevier.

[63] Alefeld, G., and Volkl, J., Eds. 1978. *Hydrogen in Metals, vol. 1, Basic Properties*. Berlin: Springer-Verlag.

[64] Wimmer, E., Wolf, W., Sticht, J., Saxe, P., Geller, C. B., Najafabadi, R., and Young, G. A. 2008. “Temperature-dependent diffusion coefficients from *ab initio*computations: Hydrogen, deuterium, and tritium in nickel.” *Physical Review B* 77:134305-134312.

[65] Yazvinskaya, N. N., Galushkin, N.E., Galushkin, D. N., and Galushkina I. A. 2018. “Investigation of hydrogen release rate from electrodes of nickel-cadmium batteries at their thermal decomposition.” *International Journal of Electrochemical Science* 13:14-22,

[66] Blanksby, S. J., and Ellison, G. B. 2003. “Bond dissociation energies of organic molecules.” *Accounts of Chemical Research* 36: 255–263.

[67] 2004. *The Hydrogen Economy: Opportunities, Costs, Barriers, and R&D Needs*, 240-250. Washington: National Academies Press.

[68] Suban, M., Tusek, J., and Uran, M. 2001. “Use of hydrogen in welding engineering in former times and today.” *Journal of Materials Processing Technology*, 119:193-198.

[69] Galushkin, N. E., Yazvinskaya, N. N., and Galushkin, D. N. 2015. “Study of Thermal Runaway Electrochemical Reactions in Alkaline Batteries.” *Journal of The Electrochemical Society* 162:A2044-A2050.

[70] Juskenas, R., Selskis, A., and Kadziauskiene, V. 1998. “In situ X-ray diffraction investigation of nickel hydride formation during cathodic charging of Ni.” *Electrochimica Acta* 43:1903-1911.

[71] Bagotsky, V. S. 2006. *Fundamentals of Electrochemistry*, 2nd ed. Hoboken: John Wiley & Sons.

[72] Vetter, K. 1967. *Electrochemical Kinetics*. New York: Academic Press.

[73] Trasatti, S. 2009. "Hydrogen Evolution." In *Encyclopedia of Electrochemical Power Sources,* edited by J. Garche, 2:41-49. Amsterdam: Elsevier.

[74] Galus, Z. 1994. *Fundamentals of Electrochemical Analysis*, edited by R.A. Chalmers, and W.A.J. Bryce. New York: Ellis Horwood.

[75] Szklarska-Smaialovska, Z., and Smaialovski, M. 1963. "Electrochemical study of the nickel-hydrogen system." *Journal of The Electrochemical Society* 110 (5):444-448.

[76] Cugnet, M., Laruelle, S., Grugeon, S., Sahut, B., Sabatier, J., Tarascon, J. M., and Oustaloup, A. 2009. "A mathematical model for the simulation of new and aged automotive lead-acid batteries." *Journal of The Electrochemical Society* 156:A974-A985.

[77] Siniard, K., Xiao, M., and Choe, S. Y. 2010. "One-dimensional dynamic modeling and validation of maintenance-free lead-acid batteries emphasizing temperature effects." *Journal of Power Sources* 195:7102-7114.

[78] Venkatraman, M., and Van Zee, J. W. 2007. "A model for the silver–zinc battery during high rates of discharge." *Journal of Power Sources* 166:537-548.

[79] Zavalis, T. G., Behm M., and Lindbergh, G. 2012. "Investigation of short-circuit scenarios in a lithium-ion battery cell." *Journal of The Electrochemical Society* 159:A848- A859.

[80] Boovaragavan, V., Methakar, R. N., Ramadesiga, V., and Subramanian, V. R. 2009. "A mathematical model of the lead-acid battery to address the effect of corrosion." *Journal of The Electrochemical Society* 156:A854-A862.

[81] Hausmann, A., and Depcik, C. 2013. "Expanding the Peukert equation for battery capacity modeling through inclusion of a temperature dependency." *Journal of Power Sources* 235:148-158.

[82] Galushkin, N. E., Yazvinskaya, N. N., and Galushkin, D. N. 2014. "Models for evaluation of capacitance of batteries." *International Journal of Electrochemical Science* 9:1911-1919.

[83] Galushkin, N. E., Yazvinskaya, N. N., Galushkin, D. N., and Galushkina I. A. 2014. "Generalized analytical models of batteries' capacitance dependence on discharge currents." *International Journal of Electrochemical Science* 9:4429-4439.

[84] Galushkin, N. E., Yazvinskaya, N. N., Galushkin, D. N., and Galushkina I. A. 2015. "Statistical models of alkaline batteries discharge." *International Journal of Electrochemical Science* 10:5530-5535.

[85] Yazvinskaya, N. N., Galushkin, N. E., and Galushkin D. N. 2018. "Analysis of generalized Peukert's equations for determining the capacity of nickel-cadmium batteries." *International Journal of Electrochemical Science* 13:8602-8614.

[86] Yazvinskaya, N. N., Galushkin, N. E., and Galushkin, D. N. 2018. "Study of discharge process final voltage influence on parameters of generalized Peukert's equation for nickel-cadmium batteries." *International Journal of Electrochemical Science* 13:9693-9700.

[87] Tremblay, O., Dessaint, L. A., and Dekkiche, A. I. 2007. "A Generic Battery Model for the Dynamic Simulation of Hybrid Electric Vehicles." Paper presented at *Vehicle Power and Propulsion Conference, VPPC 2007. IEEE*, 284-289, Arlington, September 9-12.

[88] Tremblay, O., and Dessaint, L. A. 2009. "Experimental Validation of a Battery Dynamic Model for EV Application." *World Electric Vehicle Journal* 3:1-10.

[89] Galushkin, N. E., Yazvinskaya, N. N., and Galushkin D. N. 2014. "Nonlinear structural model of the battery." *International Journal of Electrochemical Science* 9:6305-6327.

[90] Galushkin., N. E., Yazvinskaya N. N., and Galushkin. D. N. 2012. "Generalized model for self-discharge processes in alkaline batteries." *Journal of The Electrochemical Society* 159:A1315-A1317.

[91] Galushkin, N. E., Yazvinskaya, N. N., and Galushkin, D. N. 2015. "Model of relaxation processes in batteries." *ECS Electrochemistry Letters* 4:A94-A96.

[92] Galushkin, N. E., Yazvinskaya, N. N., and Galushkin, D. N. 2015. "Generalized analytical model for capacity evaluation of automotive-grade lithium batteries." *Journal of The Electrochemical Society* 162:A308-A314.

[93] Yazvinskaya, N. N., Galushkin, N. E., and Galushkin, D. N. 2018. "Analytical model of thermal runaway in alkaline batteries." *International Journal of Electrochemical Science* 13:1275-1282.

BIOGRAPHICAL SKETCHES

Nikolay Efimovich Galushkin

Affiliation: Don State Technical University, Laboratory of electrochemical and hydrogen energy, Russia.

Education: He graduated from the Southern Federal University in 1973. He received DrSc in Engineering from the South-Russian State Polytechnical University in 1998.

E-mail: galushkinne@mail.ru

Research and Professional Experience: His research interests include: First, the research and development of hydrogen storage systems meeting the criteria for on-board hydrogen storage systems that have been defined by the US Department of Energy. He received experimentally metal-hydrides high capacity. The capacity of the metal-hydrides as a hydrogen absorber was quantified as 20.1 wt% and 400 kg m^{-3}. This value exceed three times the earlier data obtained by traditional methods for any reversible metal hydrides, including magnesium hydride or complex hydrides, also they are far exceed the criteria for hydrogen storage systems established by US DOE [8] Article [14] is recognized as a key scientific article (https://advanceseng.com/chemical-engineering/thermal-runaway-new-high-performance-method-desorption-hydrogen-hydrides/).

Second, the study of the processes of thermal runaway in alkaline, acid and lithium-ion batteries. He proved by experiments that the thermal runaway in alkaline and acid batteries is connected with a powerful exothermic reaction initiation of the recombination of the atomic hydrogen accumulated in the electrodes, which runs in line with the electrochemical mechanism.

Third, the modeling of processes in electrochemical batteries to develop battery models suitable for practical use in electric vehicles.

Professional Appointments: He is a professor at Don State Technical University, Russia. He heads a research laboratory "Electrochemical and hydrogen energy."

Nataliya Nikolaevna Yazvinskaya

Affiliation: Don State Technical University, Laboratory of electrochemical and hydrogen energy, Russia.

Education: She graduated from Don State Technical University in 2004. She received Dr. Sc. in Engineering from the South-Russian State Polytechnical University in 2006.

E-mail: lionnat@mail.ru

Research and Professional Experience:

Her research interests are focused on the electrochemical processes in batteries and porous media (metal hydrides, carbon materials, nanostructures, etc.). She is an active participant in the development of hydrogen storage systems, high-capacity meeting the criteria for on-board hydrogen storage systems that have been defined by the US Department of Energy. She suggested a series of new methods of sorption and desorption of hydrogen from hydrides and carbon materials.

Professional Appointments:

She is an associate professor at Don State Technical University, Russia. She is a senior researcher in the research laboratory "Electrochemical and hydrogen energy."

Dmitriy Nikolaevich Galushkin

Affiliation: Don State Technical University, Laboratory of electrochemical and hydrogen energy, Russia.

Education: He graduated from Don State Technical University in 1998. He received DrSc in Engineering from the South-Russian State Polytechnical University in 2010.

E-mail: dmitrigall@yandex.ru

Research and Professional Experience: His research interests focus on the study of the process of thermal runaway in alkaline, acid and lithium-ion batteries. The thermal runaway is encountered in batteries of practically any electrochemical systems. Thermal runaway is a unique phenomenon in electrochemical batteries. It happens very rapidly, regardless of the attendants with the release of high energy. Despite this given phenomenon, it is very little studied. Studying and modeling of this dangerous phenomenon is extremely important for the safe operation of the plurality of electrical facilities and systems. He proved that the thermal runaway is not associated with the acceleration of known electrochemical reactions (charge batteries and the electrolyte decomposition, as thermal runaway occurs during overcharge) as it is assumed in classical mechanism of thermal runaway. He experimentally proved that the thermal runaway in alkaline and acid batteries is connected with a powerful exothermic reaction initiation of the recombination of the atomic hydrogen accumulated in the electrodes, which runs in line with the electrochemical mechanism.

Professional Appointments: He is a professor at Don State Technical University, Russia. He is a senior researcher in the research laboratory "Electrochemical and hydrogen energy."

Publications from the Last 3 Years (All authors):

1) Galushkin, N. E., Yazvinskaya, N. N., and Galushkin, D. N. 2018. “Mechanism of thermal runaway in lithium-ion cells.” *Journal of The Electrochemical Society* 165 (7):A1303-A1308.
2) Galushkin, N. E., Yazvinskaya, N. N., and Galushkin, D. N. 2018. “Preparation and Application of Metal Hydrides.” In *Hydrides: Types, Bonds and Applications*, edited by C. D. Patrick, 193-229. New York: Nova Science Publishers, Inc.
3) Galushkin, N. E., Yazvinskaya, N. N., and Galushkin, D. N. 2018. “Electrochemical methods of hydrogen absorption/desorption for metal hydrides.” In *Hydrogen Storage: Preparation, Applications and Technology*, edited by Huaiyu Shao, 197-223. New York: Nova Scince Publishers, Inc.
4) Yazvinskaya, N. N., Galushkin, N.E., Galushkin, D. N., and Galushkina I. A. 2018. “Investigation of hydrogen release rate from electrodes of nickel-cadmium batteries at their thermal decomposition.” *International Journal of Electrochemical Science* 13:14-22,
5) Yazvinskaya, N. N., Galushkin, N. E., and Galushkin D. N. 2018. “Analysis of generalized Peukert’s equations for determining the capacity of nickel-cadmium batteries.” *International Journal of Electrochemical Science* 13:8602-8614.
6) Yazvinskaya, N. N., Galushkin, N. E., and Galushkin, D. N. 2018. “Study of discharge process final voltage influence on parameters of generalized Peukert’s equation for nickel-cadmium batteries.” *International Journal of Electrochemical Science* 13:9693-9700.
7) Yazvinskaya, N. N., Galushkin, N. E., and Galushkin, D. N. 2018. “Analytical model of thermal runaway in alkaline batteries.” *International Journal of Electrochemical Science* 13:1275-1282.

8) Galushkin, N. E., Yazvinskaya, N. N., and Galushkin, D. N. 2017. "Pocket electrodes as hydrogen storage units of high-capacity." *Journal of The Electrochemical Society,* 164 (12):A2555-A2558.
9) Yazvinskaya, N. N., Galushkin, N. E., Galushkin, D. N., and Galushkina I. A. 2017. "Processes of hydrogen release relaxation at thermal decomposition of electrodes of nickel-cadmium batteries." *International Journal of Electrochemical Science* 12:2791-2797.
10) Yazvinskaya, N. N., Galushkin, N. E., Galushkin, D. N., and Galushkina I. A. 2016. "Hydrogen amount estimation in electrodes of nickel-cadmium batteries depending on their operating life." *International Journal of Electrochemical Science* 11:7843-7848.
11) 11. Galushkin, N. E., Yazvinskaya, N. N., Galushkin, D. N., and Galushkina, I. A. 2016. "Analysis of thermal runaway aftereffects in nickel-cadmium batteries." *International Journal of Electrochemical Science* 11:10287-10295
12) Yazvinskaya, N. N., Galushkin, N. E., Galushkin, D. N., and Galushkina I. A. 2016 "Study of effect of batteries capacity on probability of thermal runaway occurrence." *International Journal of Electrochemical Science* 11:8163-8168.
13) Yazvinskaya, N. N., Galushkin, N. E., Galushkin, D. N., and Galushkina I. A. 2016. "Probability investigation of thermal runaway in nickel-cadmium batteries with pocket electrodes." *International Journal of Electrochemical Science* 11:5850-5854.
14) Galushkin, N. E., Yazvinskaya, N. N., and Galushkin, D. N. 2016. "Thermal runaway as a new high-performance method of desorption of hydrogen from hydrides." *International Journal of Hydrogen Energy* 41:14813-14819.

In: Thermal Decomposition
Editor: Kalyna Jocosa Pace
ISBN: 978-1-53615-460-3

Chapter 4

A New Kinetic Model for Non-Isothermal Pyrolysis of MDF (Medium-Density Fiberboard) Derived Beech Sawdust Monitored by STA-MS Analysis: The Influence of Experimental Parameters on Process Characteristics

***Bojan Janković*[1,*] *and Nebojša Manić*[2]**

[1]University of Belgrade, Institute of Nuclear Sciences "Vinča",
Department of Physical Chemistry, Belgrade, Serbia
[2]University of Belgrade, Faculty of Mechanical Engineering,
Fuel and Combustion Laboratory, Belgrade, Serbia

---

[*] Corresponding Author's E-mail: bojan.jankovic@vinca.rs.

ABSTRACT

A novel kinetic model for pyrolysis of *MDF (medium-density fiberboard)* derived Beech sawdust was proposed. The complete procedure includes a combined four parallel reaction model (CPRM) and three-component Log-Normal distributed reactivity model. Based on the matrix analysis, it was found that for quality pyrolysis performing, the lower values of the heating rates (where the possibility of the secondary reactions is significantly reduced) are highly desirable. CPRM approach showed that with an increase of heating rate, pyrolysis rate of UF-resin adhesive rapidly increases and its rate curve moves very close to decomposition curve related to cellulose. At the highest heating rate, the UF-resin adhesive pyrolysis curve "overrated" cellulose-related curve. It was assumed that UF-resin adhesive participates in cellulose decomposition, hastening chemical bonds breaking in cellulose molecule, and further stimulating lignin decomposition. Three-component Log-Normal distributed reactivity model showed that certain asymmetries in effective activation energy distribution curves for some pseudo-components is a consequence of a wider variety of bonds present, and/or "capacity" in the tendency of an easier bond breaking. Both approaches, contained in the newly proposed model, confirm the strong deviation from the additive behavior (the adhesive compound present in the studied lignocellulose material does not produce the synergistic effect).

Keywords: *medium-density fiberboard (MDF), beech sawdust pyrolysis,* multi-component kinetic analysis, cellulose pyrolysis rate reduction, absence of synergistic effect

INTRODUCTION

In the past, the problems of production, supply, and use of energy carriers were solved within various professions. Thus, the exhaustion of certain energy resources was considered only for a certain type, or location, and completely ignored the possibility of global resource exhaustion in a relatively near future. Rational energy use is considered only within the given process through its degree of usefulness, while the decisive criterion for applying is the minimum cost of production.

Energy resources are divided into non-renewable and renewable. Inexhaustible renewable resources include radiation, hydropower, tide, waves, and wind. Other renewable resources are only conditionally inexhaustible, that is, only in the event that the rate of wear is lower than the rate of recovery. This applies to biomass or to some types of geothermal energy. Non-renewable energy resources include fossil fuels: coal, oil, natural gas, bituminous shale, etc. (Moomaw et al. 2012).

Wood industry and energy economy are closely interconnected: wood is not only the raw material of this industrial branch, but the wood is also a potential energy carrier. Remnants of industrial processing and wood processing have always been used to produce the necessary heat for the process of operation and for the heating of rooms. The primary advantage of wood as a fuel is that it is the renewable source and which we have in sufficient quantities in wood processing plants. Also, wood has a very low percentage of ash (<1%, which is much lower than in coal) and does not contain sulfur or others, in fossil fuels common, polluting and corrosive matter (Lacovidou et al. 2018). The basic components (also called *pseudo-components*) which form the structure of wood are cellulose, hemicelluloses, and lignin. The percentage share of these pseudo-components in wood substance varies from species to species, but also from the part of the wood that is analyzed. In dry hardwood, cellulose participates (as average) with 43 – 45%, lignin with 19 -26%, hexosan 3 – 6% and pentosan 21 - 26%. Within softwood, cellulose ranges between 53 – 54%, lignin 26 - 29%, hexosan 13% and pentosan 10 – 12%. Wood also contains small amounts of extractives, but they can have a great impact on the heat value of wood (Yang and Jaakkola 2011).

Wood biomass from regular logs can be used as a raw material for the production of primary and secondary fuels (pellets, briquettes, etc.), intended for combustion in the logs of boilers, from the heat generators (furnaces), in the co-generation plants, i.e., for pyrolysis processes, gasification, fermentation, etc. Pyrolysis (which represents the thermal decomposition process of carbonaceous materials in an absence of oxygen) is the first stage in the combustion and gasification of biomass in which complete or partial oxidation of primary products takes place (Bridgwater

2003). It is widely applied in chemical industry, for example, for the preparation of bio-char as a precursor for the production of activated carbon and also for cracking medium weight hydrocarbons of petroleum oil to produce lighter fractions of gasoline, etc. So, the fundamental knowledge on the thermal behavior of the pyrolysis process is the prerequisite for design and optimization of various conversion processes.

The yield of individual pyrolysis products depends on the chemical composition of the biomass and the conditions for the pyrolysis process. Pseudo-components of biomass are thermally decomposed at different rates (Demirbaş 2001), and decomposition mainly occurs by the following order: Hemicelluloses → Cellulose → Lignin. If the goal of biomass pyrolysis is to obtain the maximum yield of the liquid product, then the process is carried out at low temperatures, at high heating rates and short residence time of the biomass evaporated in the reaction zone. In order to achieve the high yield of the solid residue, the process is carried out at low temperatures and low heating rates, while for the high yield of the gas product (pyrolytic gas), the process is carried out at high temperatures, the low heating rates and long residence time of the gas phase in the reactor (Demirbaş 2001; Karaosmanoğlu et al. 1999).

In this work, the slow pyrolysis process of the waste wood material was analyzed. Namely, the pyrolysis of waste wood products is different from the one related to conventional biomass materials, because of the presence of adhesive, which can significantly influence the pyrolysis process and obtained products. Medium density fiberboard (MDF) is made using adhesives that contain urea-formaldehyde (UF) resins and is generally recognized as being highest formaldehyde-emitting pressed wood product. Therefore, the urea-formaldehyde (UF) resin that usually represents an adhesive in the chemically treated wood may influence the pyrolysis liquid products (Mu et al. 2011). The results showed (Mu et al. 2011) that there are more nitric compounds in pyrolysis liquid. However, on the other hand, there are scarce data in the literature on how this chemical compound affects the kinetics of the pyrolysis mechanism of the waste wood and the impact on the bio-char yield and the rate of the development of the gaseous components (Lai et al. 2018; Li et al. 2012; Li et al. 2014; Chen et al. 2015). This particularly applies to the effect of adhesive on the change of lignocellulosic material with particular reference to the cellulose structure. This refers to the development of a new and simple kinetic model that could explain these transformations.

The main goal of this chapter is proposal of a new kinetic model that describes the non-isothermal pyrolysis process of MDF derived Beech sawdust samples, which is monitored by simultaneous thermal analysis (STA) (STA includes TGA (Thermogravimetric analysis) – DSC (Differential scanning calorimetry) techniques) coupled with Mass spectrometry (MS) analysis of the gaseous products. The proposed kinetic model introduces the linear combination of the Log-Normal distributions in the framework of the distributed effective activation energy model (DAEM). Also, the influence of the process parameters, such as the heating rate, time and temperature on the behavior of the pyrolysis conversion curves was investigated through the application of a matrix analysis.

EXPERIMENTAL PROCEDURE

Material and Methods

The MDF Beech sawdust was obtained from local carpenter workshop in a plastic transparent bag of 1 kg, where, according to the delivery from the supplier, it is noted that MDF wood waste was treated in accordance with requirements of the EN standard for MDF (EN 622-5 2006).

According to standard EN ISO 14780 (EN ISO 14780 2017) for solid biomass sample preparation, the sample was prepared for the experimental tests. A sample was removed from the transportation packing, and then it is placed inside the oven at 105 °C for 2 h, in order to define moisture left on the inner surfaces of the packing. This amount of moisture was included in the calculation of the total moisture content in the sample according to standard EN ISO 18134 (EN ISO 18134 2015). Further, regarding pre-drying process of the sample, which is necessary to remove the residual moisture during preparation, the sample was placed on a plate and left to reach moisture equilibrium with laboratory atmosphere conditions for 24 h. After that period of time, the sample was reduced in two steps using a cutting mill, after which it was sieved through 1 mm (18 mesh sizes) sieve.

Finally, the obtained undersize was declared as a general analysis sample and divided into four test portions for further testing. The first portion of the prepared sample was used for determination of ultimate analysis (carbon, hydrogen and nitrogen content) in accordance with EN ISO 16948 (EN ISO 16948 2015). The results of the proximate analysis (total moisture, ash, volatile matter, and char content) were obtained from the second portion of the prepared sample in accordance with EN ISO 17225-1 (EN ISO 17225-1 2014). The higher heating value (HHV) and lower heating value (LHV) for the tested sample were determined using calorimeter laboratory equipment with the oxygen bomb (IKA C200, IKA® Works, Inc., Wilmington, USA), according to the standard EN ISO 18125 (EN ISO 18125 2017). The third portion was used for determination of the lignocellulosic content of the material. The content of cellulose in the sample was determined by the Kürschner-Hoffer's method (Schuerch 1967), while lignin content after extraction (toluene-ethanol) was determined by the Klason's method (TAPPI 222 om-02 2006) with the spectrophotometric determination (Specord® Plus UV/Vis Spectrophotometer, Analytik Jena AG, Analytical Instrumentation Konrad-Zuse-Str. 1, 07745 Jena, Germany) of acid-soluble lignin (based on absorption of ultraviolet radiation, with most often used wavelength of 205 nm) according to TAPPI method. The content of hemicelluloses is determined approximately, as a complement to the content of certain components up to 100%. The results are expressed in relation to the absolute dry weight of the studied material and presented as the mean (arithmetic mean) for the four repeated measurements.

The STA-MS was performed on the fourth portion of the prepared biomass sample, and all of the obtained results were used for studying the pyrolytic kinetics.

NETZSCH STA 445 F5 Jupiter system was used for STA experimental tests for examined samples. The inert atmosphere was provided to maintain the pyrolysis process using high purity nitrogen

(Class 4.6) as a carrier gas. At the same time, nitrogen was used as a protective gas in order to keep the high sensitive internal balance (0.1 µg). Both carrier and protective gas flow were set to φ = 50 mL min^{-1}. The weight measurements were carried out using internal balance which provided the following result as 5.70 ± 0.10 mg of testing sample. Crucibles that were used for tests are made of alumina, and during each measurement, they were filled approximately up to the half. Crucible's lid was placed on the top, so the optimum heat transfer could be realized. Each sample was tested using three different heating rates, one for each measurement. It was considered that the heating rates of β = 5, 10 and 20°C min^{-1} seems like the best optimized rates for kinetic data collection. Using these heating rates, the tested samples were heated from room temperature up to 600 °C.

During all measurements, the sample temperature controller (STC) was turned off, so the set temperature (600°C) is referred to the furnace temperature. Temperatures presented on diagrams are the sample temperatures that, due to the construction of the furnace, never reached the set temperature. This fact resulted in better temperature curve linearity than it would be if STC was turned on. The STA 445 F5 Jupiter runs under the versatile Proteus® software and includes all operating tools to obtain a reliable measurement and evaluate the resulting data, or even carry out complicated analyses.

Simultaneously with TGA testing, the DSC (Differential scanning calorimetry) recording was performed in the furnace of the NETZSCH STA 445 F5 Jupiter system for the purpose of precise determination of the specific heat and caloric effects. The DSC instrument works in accordance with the heat-flux principle and features the high detection sensitivity. Obtained results from these measurements were exported and prepared for further kinetic analysis.

Determination of the evolved gases from performed STA was carried out continuously using the quadrupole mass spectrometer NETZSCH QMS 403 D Aëolos (QMS) (Erich NETZSCH GmbH & Co. Holding KG, Germany). The STA – QMS coupling was done using transfer line with quartz capillary tube with a diameter of 75 µm. The whole transfer line was heated up to 230 °C in order to avoid condensation of evolving volatile compounds. The QMS was operated under vacuum up to the 10^{-7} bar, providing conditions necessary to detect gas components using their ions intensity according to their respective mass to charge ratios (*m/z*). Evolved gas composition was monitored and analysed through bar-graph cycles, scanning mass units in the range from 1 to 80. Cycles were set to speed of 0.2 s per mass unit and carried out using stair mode. The excitation energy in the QMS was set up at 1200 eV with the resolution of 50 units.

From the selected range, the focus was on specific ions that correspond to gases evolved in the pyrolysis process. Accordingly, the molecules with atomic mass units (amu) of 2, 16, 18, 28, 30, 44, 58 and 72 which correspond to H_2, CH_4, H_2O, CO, C_2H_6, C_3H_8 (CO_2), C_4H_{10} and C_5H_{12} respectively, were analysed, and the intensity peak areas obtained for each compound were compared.

KINETIC MODEL FOR PYROLYSIS PROCESS

During TG-DSC measurements, the considered biomass sample decomposed under program-controlled heating in accordance with the relation: $T = T_o + \beta \cdot t$, where T_o is the starting operation temperature (usually ambient), β is the applied heating rate [°C min^{-1}], while t is the time [min]. The general rate-law equation which describes the loss of the mass of biomass sample during controlled heating can be presented in the form:

$$\beta \cdot \left(\frac{d\alpha}{dT} \right) \equiv \left(\frac{d\alpha}{dt} \right) = A \cdot \exp\left(-\frac{E_a}{RT} \right) \cdot f(\alpha), \quad (1)$$

where $\beta = \mathrm{d}T/\mathrm{d}t$ = const. is the heating rate, α is the conversion fraction of starting material which is dimensionless ($\alpha = (m_o - m_T)/(m_o - m_\infty)$ where m_o is the initial mass of the sample, m_T is the current mass of the sample at temperature T and time t, while m_∞ is the residue mass of pyrolysis at the end of the process), A is the pre-exponential factor [min^{-1}], E_a is the effective (apparent) activation energy, R is the gas constant (8.314 J mol^{-1} K^{-1}) and T is the absolute temperature; $f(\alpha)$ represents the function related to conversion fraction α, whose format depends on the actual reaction mechanism of the pyrolysis. Under the non-isothermal conditions, by reorganizing the Eq. (1) and through integration procedure, we obtain:

$$\int_0^{\alpha} \frac{d\alpha}{f(\alpha)} = g(\alpha) = \frac{A}{\beta}\int_0^{T} \exp\left(-\frac{E_a}{RT}\right) dT \cong \left(\frac{AE}{\beta R}\right) \cdot p(x), \quad (2)$$

where $g(\alpha)$ is the integral form of reaction mechanism function, $T_o << T$, so $T_o \rightarrow 0$ K, while $p(x)$ represents the temperature integral as:

$$p(x) = -\int_{\infty}^{x} \left(\frac{\exp(-x)}{x^2}\right) dx, \quad (3)$$

where $x = E_a/RT$ is the reduced effective activation energy. Namely, $p(x)$ has no exact analytical solution, so various approximations exist for $p(x)$, such as Doyle's (Doyle 1962) approximation, or the Senum and Yang's (Pérez-Maqueda and Criado 2000) fourth degree approximation which give sufficiently reliable results.

In order to obtain the reaction order (n) in the case of non-isothermal data, where the variation of α with T and β can be described by equation (Avrami 1939):

$$\alpha(T) = 1 - \exp\left[\frac{k(T)}{\beta^n}\right], \quad (4)$$

where $k(T) = A \cdot \exp(-E_a/RT)$ is the Arrhenius temperature-dependent rate constant, taking the double logarithm of the both sides of above-equation, with substituting Arrhenius relation, we finally obtain:

$$\ln\left[-\ln\left(1-\alpha\left(T\right)\right)\right]=\ln A-\frac{E_a}{RT}-n\ln\beta. \tag{5}$$

From the plot of ln[- ln(1 - α(T))] against lnβ, which can be obtained at the same temperature from a number of the isotherms considered at various heating rates, the slope of the estimated straight line gives the value of n. It should be noted that n is still called the Flynn-Wall-Ozawa exponent (Ozawa 1965).

The increase in the accuracy of kinetic analysis can be reached by application of isoconversional methods which compute the temperature integral as an integral part of the evaluating the values of $E_{a,\alpha}$ (the particular E_a value at selected α). One of these methods is Vyazovkin isoconversional method (Vyazovkin 2001) that is very flexible and does not introduce the systematic error in the computational procedure for E_a evaluation and can be applied to any one of the heating programs. So, the E_a value can be determined by minimizing the following equation:

$$\Psi\left(E_{a,\alpha}\right)=\sum_{i=1}^{N}\sum_{j\neq i}^{N}\frac{J\left[E_{a,\alpha},T_i\left(t_\alpha\right)\right]}{J\left[E_{a,\alpha},T_j\left(t_\alpha\right)\right]}, \tag{6}$$

where

$$J\left[E_{a,\alpha},T\left(t_\alpha\right)\right]=\int_{t_\alpha-\Delta\alpha}^{t_\alpha}\exp\left[-\frac{E_{a,\alpha}}{RT\left(t\right)}\right]dt. \tag{7}$$

The $T(t)$ is the set of arbitrary temperature programs, while N represents different heating rates. Integral is solved numerically by minimization of $\Psi(E_{a,\alpha})$ where the minimization is repeated for each value of α in order to obtain the dependence of $E_{a,\alpha}$ on α. The Eq. (7) allows substitution of actual sample temperatures for $T(t)$, so the use of actual sample temperature in the Eq. (7) conjuration the current method applicable to any present program (i.e., non-isothermal or isothermal, or for the linear heating or not). The systematic error can be eliminated if the integration was performed over small time segments, so the $E_{a,\alpha}$ value is assumed to be constant only within a small interval of $\Delta\alpha$ which is typically taken to be 0.01 – 0.02. In other integral isoconversional methods, as a result of mentioned integration, the calculated $E_{a,\alpha}$ values are accurate only when $E_{a,\alpha}$ does not vary with α. Differently, $E_{c,\alpha}$ estimates contain the systematic error whose size increase with the increasing range of E_a variability. This error can be avoided by application of the above-stated approach. For the studied pyrolysis process, the step of $\Delta\alpha = 0.01$ was used. The confidence interval for effective activation energies can be calculated using the appropriate statistical approach described (*Vyazovkin* and *Sbirrazzuoli 1997*).

From obtained isoconversional results, the resulting E_a values at various α's can be used as the initial parameter values in the fitting process and as a comparison for the final kinetic results. The isoconversional approach can be useful to give confidence that the estimated kinetic parameters and employed reaction model (for example, the first or the *n*-th order) are reliable representations of the chemical process which takes place.

The Combined Four Parallel Reaction Model (CPRM)

Pyrolysis process of lignocellulosic material such as studied system can be assumed to follow four parallel reaction models (see Scheme 1)).

$$Sawdust - MDF \rightarrow \begin{cases} Hemicelluloses \rightarrow volatiles + bio - char \\ Cellulose \rightarrow volatiles + bio - char \\ Lignin \rightarrow volatiles + bio - char \\ Adhesive \rightarrow volatiles + bio - char \end{cases}$$

Scheme 1. The combined four parallel reaction model (CPRM) for MDF derived Beech sawdust pyrolysis.

In Scheme 1, it can be assumed that the adhesive decomposition process may affect on reaction mechanism pathway alterations for other decomposition reactions of three pseudo-components.

Thermal decomposition of *MDF* derived Beech sawdust can be modeled by assuming pseudo-components (hemicelluloses, cellulose, and lignin) decomposition reactions together with adhesive specie decomposition, through combined four parallel *n*-th order reactions, such as:

$$\frac{d\alpha_l}{dt} = \varphi_l A_l \cdot \exp\left(-\frac{E_{a,l}}{RT}\right) \cdot \left(1-\alpha_l\right)^{n_l}, \tag{8}$$

and

$$\frac{d\alpha_M}{dt} = \sum_{l=1}^{M} \frac{d\alpha_l}{dt}, \qquad \alpha_M = \sum_{l=1}^{M} \alpha_l, \tag{9}$$

where $d\alpha_l/dt$ is the rate of *l*-th component decomposition, A_l is pre-exponential factor of *l*-th component, while $E_{a,l}$ and n_l represent apparent activation energy and reaction order related to the decomposition of *l*-th component, respectively; α_l is the conversion fraction values within the range where *l*-th component decomposes, α_M represents the total conversion fraction that composes all α_l values for *M* reactions present, while φ_l is the contribution of the partial reaction to the overall process (the definition of this parameter is dependent on the concentration or mass fraction of the *l*-th component in the studied sample).

The kinetic parameters φ_l, A_l, $E_{a,l}$ and n_l for each component during pyrolysis can be obtained by minimizing the objective function (OBF) described by the following equation:

$$OBF = \sum_{k=1}^{N_k} \sum_{m=1}^{N_m} \left[\left(\frac{d\alpha_M}{dt}\right)_{\exp,k,m} - \left(\frac{d\alpha_M}{dt}\right)_{calc,k,m} \right], \tag{10}$$

where N_k and N_m are the number of heating rates and the numbers of TGA measured, respectively.

The procedure was performed using the multi-dimensional non-linear regression function *lsqcurvefit* in Matlab R2016b. This function is commonly used for solving nonlinear curve-fitting (data-fitting) problems in the least-squares sense. Conversely, minimizing Eq. (10) requires a mathematical evaluation of the Eq. (9) for all experimental temperature ranges and iteration process which was performed by *while loop* in Matlab R2016b. The model quality of the fit is determined by Eq. (11) for each overall simulation process and their heating rates:

$$QFit = 100 \times \sum_{m=1}^{N_m} \frac{\sqrt{\dfrac{\left[\left(\dfrac{d\alpha_M}{dt}\right)_{\exp,m} - \left(\dfrac{d\alpha_M}{dt}\right)_{calc,m}\right]^2}{N_m}}}{\max\left|\dfrac{d\alpha}{dt}\right|_{\exp,m}}, \tag{11}$$

where m is the number of data points, max $|d\alpha/dt|_{\exp,m}$ is the maximum experimental value (the highest value of decomposition rate registered during the experiment).

Considering the mathematical formulation of the multicomponent kinetic modeling, in order to solve all presented equations, defining the first prediction of all kinetic parameters φ_i, A_i, $E_{a,i}$ and n_i for each component was necessary. Values for the first prediction were obtained according to the results of isoconversional approach and were used for the start of the iteration process. In each iteration step, *lsqcurvefit* function for solving nonlinear curve-fitting was set to use the *trust-region-reflectiv* algorithm, with function and step tolerance equals to 10^{-12} and the maximum function evaluations equal to 10^3. The criteria for the final solution of the calculations according to results of the kinetic parameters for each pseudo-component were the minimum value of *QFit*.

In order to obtain information of an additive effect of the adhesive compound in a complex heterogeneous system which consists of three main pseudo-components such as hemicelluloses ("1"), cellulose ("2") and lignin ("3"), the synergy effective activation energy ratio (SEAER) can be presented as:

$$SEAER = \frac{ACE_a^{\,*}}{0.5\times(E_a^{(1)})+0.5\times\left(E_a^{(2)}\right)+0.5\times\left(E_a^{(3)}\right)}, \tag{12}$$

where ACE_a^* is the effective activation energy of the adhesive degradation, while $E_a^{(1)}$, $E_a^{(2)}$ and $E_a^{(3)}$ represent the effective activation energy values related to hemicelluloses, cellulose, and lignin decomposition processes during pyrolysis, respectively. The constant 0.5 is introduced as equimolar constants that are initially associated with each observed pseudo-component. If SEAER > 1, this means that effective activation energy of the adhesive is much higher than the expected E_a in the system with other three components (pseudo-components of the biomass). If SEAER = 1, it means that effective activation energy of the adhesive is approximately equal to the expected E_a with the entire biomass system during pyrolysis. If SEAER < 1, this means that effective activation energy of the adhesive is lower than the expected E_a in the system with other three components.

Distributed Reactivity Model Consisting the Linear Combination of the Log-Normal Distributions of Effective Activation Energies

The distributed reactivity model (DRM) of the effective activation energy counterparts represents multiple parallel reaction processes which can involve complex pyrolysis mechanism of solid biomass sample. If the E_a (and corresponding A) and the reaction model together with the contribution of each parallel reaction is determined, the kinetic modeling of the pyrolysis process can be further performed through DRM approach. The reaction order models with various n can be directly used for simulation the pyrolysis behaviors of each component in the studied reaction system. In order to estimate the distribution of the effective activation energies of the entire pyrolysis process, the Log-Normal distribution $f_p(E_a)$ ($p = 1, 2, \ldots$) with the mean value of μ_p [kJ mol^{-1}] and standard deviation σ_p [kJ mol^{-1}] was assumed, as:

$$f_p\left(E_{a,p}\right)=\frac{f_{0,p}}{\sqrt{2\pi}\sigma_p\cdot E_{a,p}}\exp\left\{-\frac{\left[\ln\left(\frac{E_{a,p}}{\mu_p}\right)\right]^2}{2\sigma_p^{\,2}}\right\}, \tag{13}$$

where $f_{0,p}$ represents the area under the distribution curve related to p-component, which can proportionally show the participation of individual components in bond breaking degree capacities.

In the case of n-th order reaction models, the overall conversion fraction equation which encompasses all components of the biomass sample during pyrolysis can be expressed as:

$$\alpha=1-\left\{1-\left(1-n_p\right)\cdot\int_0^{\infty}\exp\left[-\frac{A_p}{\beta}\int_0^T\exp\left(-\frac{E_{a,p}}{RT}\right)dT\right]\right\}^{\frac{1}{(1-n_p)}}\cdot \tag{14}$$
$$\cdot\left[\sum_p w_p f_p\left(E_{a,p}\right)\right]\cdot dE_{a,p},$$

where

$$w_1+w_2+\ldots+w_p=1,\quad \sum_p w_p=1, \tag{15}$$

whereby w is the weighting factor, that scales the devolatilization contributions from the parallel reactions. Double Exp-function (DEF) which exists in the Eq. (14) as the $DEF=\exp\left[-\frac{A}{\beta}\int_0^T\exp\left(-\frac{E_a}{RT}\right)dT\right]$ can be solved by the application of the Gauss-Hermite quadrature numerical integration (Farias et al. 2013). The DEF depends upon temperature condition used in the conducted experiment, and the initial distribution estimated may depend on the physical behavior of the studied biomass system.

The pattern search algorithm (Pattern Search Method (PSM)) is employed in order to perform the calculation process for determination the kinetic parameters involving calculated processing variables, such as the pyrolysis rates ($(\mathrm{d}\alpha(A_p,E_{a,p},\mu_p,\sigma_p)/\mathrm{d}t)_{calc}$), at every considered heating rate, and comparing them with experimental ones.

RESULTS AND DISCUSSION

Ultimate/Proximate Analysis and Lignocellulosic Content

Table 1 shows the results of the ultimate/proximate analysis and lignocellulosic contents of MDF (the medium-density fiberboard) derived Beech sawdust used in the current investigation, together with the data of the other wood species for the purpose of comparative analysis.

It can be observed that MDF Beech sawdust used in this work is characterized with high volatile matter content (79.62%), which is also evident in the case of VM for Pinewood chips, Pine sawdust and Poplar wood (Table 1). This is important for thermal properties of wood related to combustion or gasification processes. Because of these characteristics, the studied system is desirable for a good regulation of combustion processing. Also, the volatile matter may impact on the realizing yield of light hydrocarbons (primarily CH_4), CO, CO_2, H_2 and formation of tar. In addition, the higher volatile matter of investigated system (together with similar systems with the same property (VM) shown in Table 1) may result in the highly porous bio-char, and accelerating the bio-char combustion. This property also can be attached to the highest conversion, in comparison to wood sample with the highest fixed carbon (Pine bark, Table 1).

Table 1. The ultimate and proximate analysis results and lignocellulosic content of MDF derived Beech sawdust sample used in actual work, together with data of other wood species for the purpose of comparative analysis (data collected from available scientific) (adhesive analysis is not included)

Sample	Reference	Ultimate analysis (wt %)					Proximate analysis (wt %)				HHV ($MJ\ kg^{-1}$)	LHV ($MJ\ kg^{-1}$)	Lignocellulosic content		
		C	H	N	S	O	Moisture	Volatile matter (VM)	Fixed carbon (FC)	Ash			Cellulose (%)	Hemicelluloses (%)	Lignin (%)
MDF Beech sawdust[a]	This work	48.63	6.55	2.01	0.00	42.20[c]	5.40	79.62	14.40	0.58	18.15	16.62[b]	41.70	34.05	24.25
Wood sawdust	(Cao et al. 2006; Joshua et al. 2016)	46.20	5.10	1.50	0.06	35.40	10.40	70.40	17.90	1.30	-	16.22	41.03	39.95	19.02
Particle board	(Chen et al. 2011; Lamaming et al. 2013)	45.01	6.25	0.09	0.01[d]	40.94	7.45	75.08	17.22	0.25	-	16.45	59.90	9.90	21.00
Eucalyptus wood	(Channiwala and Parikh 2002; Pereira et al. 2013)	38.49	4.86	0.25	0.00	37.19	16.40	62.99	17.81	2.81	-	12.98	47.60	22.10	30.30
Forest residues	(Vamvuka et al. 2003; Burkhardt 2013)	49.27	5.72	0.28	0.08	37.15	7.32	73.96	18.54	0.18	-	17.83	49.80	14.10	36.10
Wood chips - Pine	(Phanphanich and Mani 2011; Cotana et al. 2014)	44.05	6.20	0.16	0.00	42.65	6.69	80.23	12.84	0.25	-	15.82	32.09	14.22	31.15
Wood residue	(Senneca 2007; Nanda et al. 2013)	43.53	5.71	0.39	0.00	40.80	7.94	75.65	14.79	1.63	-	15.30	48.30	22.10	22.60
Wood chips	(Senneca 2007; Foster et al. 2010)	30.79	3.97	0.13	0.00	29.95[d]	34.90	51.60	13.30	0.26	-	9.84	41.00	30.00	21.00

Table 1. (Continued)

Sample	Reference	Ultimate analysis (wt %)					Proximate analysis (wt %)				HHV (MJ kg^{-1})	LHV (MJ kg^{-1})	Lignocellulosic content		
		C	H	N	S	O	Moisture	Volatile matter (VM)	Fixed carbon (FC)	Ash			Cellulose (%)	Hemicelluloses (%)	Lignin (%)
Pine bark	(Şensöz 2003; Saarela et al. 2005)	48.02	5.17	0.43	0.00	42.12[d]	3.00	69.84	25.87	1.26	-	16.10	22.20	8.10	13.10
Oak	(Şensöz 2003; Isikgor and Becer 2015)	47.06	5.67	0.28	0.09	39.90	6.50	73.00	20.00	0.50	-	16.63	40.40	35.90	24.10
Oak sawdust	(Vassilev 2010; Aigbomian and Fan 2013)	44.19	5.20	0.09	0.01	38.72	11.50	76.30	11.90	0.30	-	15.13	42.00	29.40	28.60
Pine sawdust	(Vassilev 2010; Tarrés et al. 2017)	43.15	5.08	0.08	0.01	36.28	15.30	70.40	14.20	0.10	-	14.91	55.06	12.30	30.60
Pine sawdust	(Mishra and Mohanty 2018)	50.30	6.00	0.69	0.00	42.99	6.09	78.03	12.16	2.07	18.44	-	55.92	15.35	10.55
Fir wood	(Vassilev 2010; Isikgor and Becer 2015)	48.53	5.66	0.28	0.09	38.24	6.70	75.70	17.10	0.50	-	17.36	44.00	11.00	27.00
Poplar wood	(Vassilev 2010; Isikgor and Becer 2015)	47.06	5.56	0.55	0.02	38.01	6.80	79.70	11.50	2.00	-	16.77	50.80-53.30	26.20-28.70	15.50-16.30
Willow	(Vassilev 2010; Bodîrlău et al. 2007)	44.07	5.40	0.53	0.05	38.45	10.10	74.20	14.30	1.40	-	15.41	46.76	24.60	28.64
Sal sawdust	(Mishra and Mohanty 2018)	49.83	6.01	0.58	0.00	43.56	8.88	76.03	14.09	1.14	18.20	-	52.36	14.59	11.18

[a] Ultimate analysis: On a dry basis.

[b] EN ISO 18125 2017.

[c] By the difference.

[d] It was obtained by calculating.

Among the compared systems, the MDF Beech sawdust belongs to the medium ash giving system (Table 1). The content of ash is linked with a used particle size of the material, where the smaller particle size can produce more *ash content* compared to larger particle size (Aqsha et al. 2011). The lower ash content would be desirable because it, could negatively affect the yields to partially eliminate the formation of bio-char (Pereira et al. 2014). Therefore, taking into consideration our studied system, the ash content is below 1%, which is exceptionally favorable, for example, the production of activated carbons.

The raw material has a moisture content of 5.40% which can be attributed to the open air conditions as well as to wet separation process. From the results of proximate analysis of the tested sample compared with other biomasses, the MDF Beech sawdust it could be suitable for the good production of bio-oil. On the other hand, the obtained HHV value for the MDF Beech sawdust (18.15 MJ kg^{-1}) belongs to the heating values of biomass derived bio-chars (11.83 – 44.20%) (Table 1) (Mohan et al. 2006).

The resulting LHV value for the MDF Beech sawdust is typical for the wood and wood waste products (Table 1), there exists a good balance between HHV and LHV (including the relationship between ash and moisture). Namely, the average calorific value of wood and wood products depends on the density of the wood, whereby the density of the wood is affected by the type and age of the wood, the part of the wood being analyzed (trees, branches, root, leaves, etc.), as well as the timing of wood cutting (Table 1).

The MDF derived Beech sawdust is characterized by the higher concentrations of C and H in accordance with the values of other types of wood sawdust (Table 1). The contribution values of C, H, and O in the studied system are within the specified range for the considered element (Table 1), while the lower content of N (2.01%) identified for MDF derived Beech sawdust originates from the presence of an adhesive substance in the sample. Taking this into account, the main product of nitrogen transformation during the pyrolysis would be NH_3, but it should be noted that application of the higher heating rates may significantly mitigate the rate of this transformation and reduce the amount of NH_3.

With reference to the lignocellulosic content, the MDF derived Beech sawdust has the higher cellulose percentage (41.70%) and the lower content of lignin (24.25%) (Table 1). The extent of the cellulose content is typical for wood sawdust systems (Table 1) and the value is in the range of percentages for wood biomass. It was found that the percentages of cellulose and hemicelluloses for MDF derived Beech sawdust is in good agreement with a balanced ration of C and O carriers in the lignocellulosic material, but the lignin consists of several different carbon compounds, where the carbon represents 60% of the mass of the lignin. However, the content of cellulose, hemicelluloses and lignin may vary depending on the type of the biomass (Stefanidis et al. 2014).

In accordance with the above discussion, it should be pointed out that an important parameter, for example, the obtaining a quality activated carbons production, is the contents of cellulose and lignin in the raw material. Comparing our system with others presented in Table 1, it can be observed that great variety in cellulose and lignin contents exists, where MDF derived Beech sawdust belongs to those biomasses with increased cellulose content.

Such differences can be attributed to different procedures of extraction, where the extraction method plays an important role. However, for MDF derived Beech sawdust, the cellulose content exceeds the content of lignin, so we can assume that systems with high cellulose content may produce the bio-char with pronounced microstructure (Table 1) (Savova et al. 2001). This is followed with the high present share of carbon that exceeds 40% (Table 1), so therefore, the MDF derived Beech sawdust represents a good precursor for the activated carbons.

The high C and O contents followed with the high cellulose percentages are the optimal parameters for the prediction of heating values of biomass and they are the main contributors of the energy content in considered biomass. The presence of these factors in the lignocellulosic material leads to a more bio-char formation, as well as the high calorific values (Table 1). In contrast, the nitrogen (also, the sulfur content dropped to zero) content is very small, that additionally supports evidence that the pyrolysis of the MDF derived Beech sawdust does not result in the release of the noxious gases.

STA Analysis of MDF Beech Sawdust Pyrolysis

Figure 1 a) and b) shows experimental TG (Thermogravimetric) – DSC (Differential scanning calorimetry) curves for thermal decomposition of MDF derived Beech sawdust, at the various heating rates in a nitrogen atmosphere. It can be observed (Figure 1 a)) that the moisture was removed from raw material up to 175°C, and the main decomposition reactions start around 200°C and finished around 590°C. The pre-stage belongs to the dehydration reaction and the loss of water molecules from the sample, which is characterized by the mass loss of approximately 6.6% (Figure 1 a)). After this pre-stage, it follows the main decomposition stage, which represents the devolatilization stage, encompasses realizing the vapors and gases. The actual stage is characterized by the great mass loss of the tested sample (~ 62.3% (Figure 1 a))). This process segment represents the active pyrolysis zone.

The devolatilization is a key step in the thermo-chemical conversion of biomass and describes the process where the volatile matter which consisting the gases and condensates are released as the solid fuel is heated in an inert atmosphere (Figure 1 a)). A release of volatiles is the consequence of the scission of chemical bonds in biopolymers which comprises the lignocellulosic material. Consequently, the reaction rate, the reaction order, and the product yields depend on the important parameters, such as the temperature, the heating rate, pre-treatment, and additional catalytic effects (Bridgwater et al. 2002). Pyrolysis mechanism can be approximated through the combination of the contributions from three pseudo-components which are dominant in the lignocellulosic materials. In this respect, the cellulose is in a focus of the investigation, since, it is a dominant lignocellulosic compound. If we take a look at Figure 1 a), the devolatilization stage includes primary decomposition of hemicelluloses and cellulose, but the prominent mass loss during the pyrolysis process is probably caused by the decomposition of cellulose.

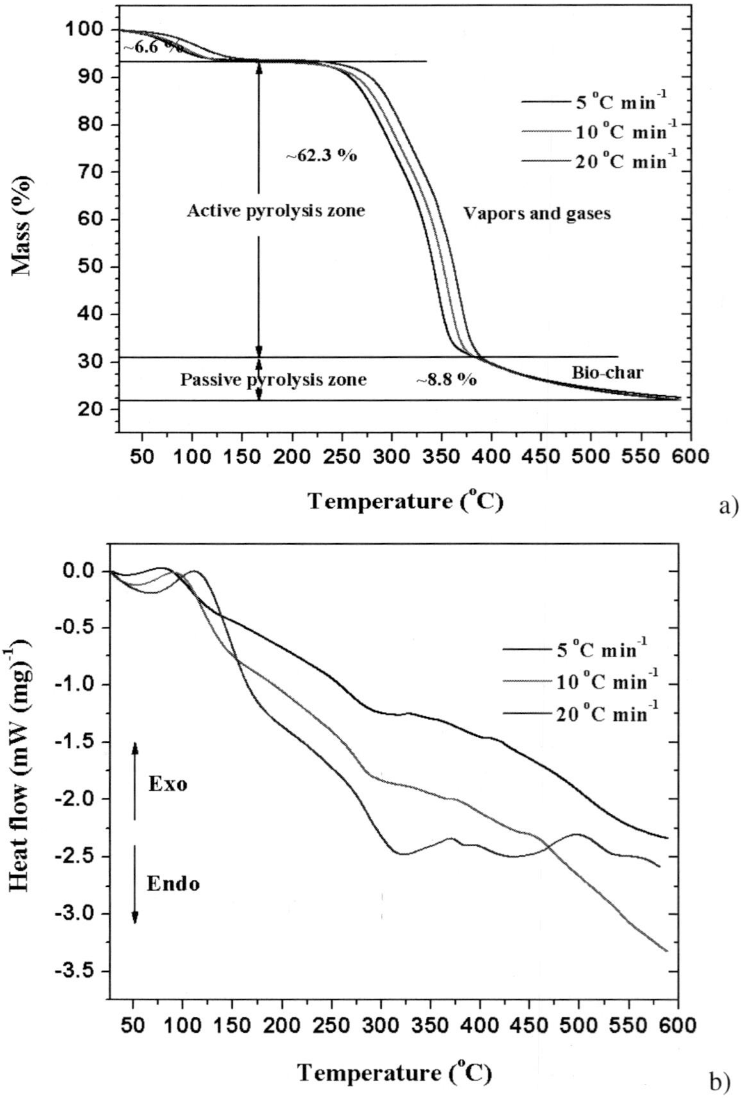

Figure 1. TG - (Thermogravimetric) a) and DSC - (Differential scanning calorimetry) b) curves for thermal decomposition of MDF derived Beech sawdust, at the various heating rates in a nitrogen atmosphere.

Since the hemicelluloses contains hetero-polysaccharide, unlike the cellulose which has only glucose in its molecular chain structure, the hemicelluloses starts to decompose much earlier than cellulose, in the range of 175 – 350°C. The different content of cellulose in various biomasses varies depending on its crystallinity, which can impact on the bio-char yield and also on the reaction mechanism. The decomposition of hemicelluloses was immediately followed by cellulose decomposition at higher temperatures, in the range of 275 – 350°C (Basu 2010), in view of its greater thermal stability from hemicelluloses, even when considered in a pure state, independently as an integral part of the lignocellulosic material. It can be noted that the thermal stability of individual pseudo-components depends on the temperature levels applied, closely connected to the heating mode (fast or slow). The lignin exercises intermediate thermal decomposition characteristics (260 – 500°C) but in a much wider temperature region. The major expected chemicals which can be obtained from hemicelluloses pyrolysis are the acetic acid and furfural, while major expected chemicals obtained from cellulose pyrolysis are primarily levoglucosan and hydroxyacetaldehyde. These compounds can be obtained right within the temperature range, which belongs to the active pyrolysis zone shown in Figure 1 a). From the lignin decomposition in a wide temperature interval (approximately 175 – 600°C), the most value added chemicals are phenolic compounds and oxygenate with C_1 – C_{15} (Bridgwater 2009). The passive pyrolysis zone corresponding to the mass loss of 8.8% (Figure 1 a)) can be attributed to bio-char formation with existing a rigid amorphous carbon matrix solid.

The thermal heat effects during the pyrolysis of MDF derived Beech sawdust can be viewed on corresponding DSC curves at the various heating rates (Figure 1 b)). It can be observed that the heat effect varies with a changing the heating rate. A fairly deep endothermic peak extending from room temperature up to 130°C can be attributed to water evaporation, while endothermic effects in the wider temperature range from 250 to 375°C at all heating rates, can be linked with the decomposition reactions of hemicelluloses and cellulose. It should be noted that the "strength" of this endothermic phenomenon varies with the application of the heating rate, and increases with the increase of the rate of heating (with the effect of shifting to higher temperatures, typically for the thermally induced transformations). At the temperatures higher than 400°C, it can be observed slight (except for β = 20°C min^{-1}) exothermic effect (within the passive pyrolysis zone (Figure 1 a))) which can be attributed to the bio-char forming. Namely, the certain variations in the heat effects can be significantly stimulated by a concentration of three pseudo-components (hemicelluloses, cellulose, and lignin) present in the sample, which also affects the development of their decomposition paths.

Figure 2 shows DTG (derivative thermogravimetry) curves of the pyrolysis process of MDF derived Beech sawdust, recorded at various heating rates (5, 10 and 20°C min^{-1}) in an inert (N_2) atmosphere. DTG curves of MDF derived Beech sawdust pyrolysis clearly shows three distinct regions, attributed to the following processes: the water removal from the material occurring up to ≈ 175°C, the active pyrolysis zone where devolatilization reactions take place (ΔT = 195 – 410°C), and finally, the passive pyrolysis zone (for T > 410°C) related primarily to bio-char formation process, respectively. However, within the devolatilization stage, it can be observed the formation of DTG "shoulder" and one main DTG peak, positioned in the temperature ranges of ΔT_{sh} = 290 – 320°C, and ΔT_{peak} = 345 – 368°C, respectively, depending on the applied heating rate. The formation of "well-defined peak" or "shoulder" in the first indicated temperature region can be attributed to the decomposition of hemicelluloses, but this depends on the hemicelluloses content in the raw material. In accordance with the analysis presented in Table 1, the hemicelluloses content was typical for hardwoods, since its content is higher in hardwood than in softwood. The decomposition of hemicelulose is faster in hardwood species than in softwood species, where Grønli et al. (2002) observed that the hemicelluloses decomposition is slower in softwoods than in hardwoods by studying softwoods from pine species and several hardwood species. The hemicelluloses reactivity increased in hardwood than in softwood, which can be related to differences in the chemical composition of two species hemicelluloses (Grønli et al. 2002). The well-defined peaks in the temperature range of 345 – 368°C at all observed heating rates can be attributed to cellulose decomposition reactions. It can be observed that depending on the heating rate applied, the 'intensity' of the mass loss rate attributed to cellulose decomposition varied.

Namely, it was found (Aviles and Brown 2014) that heating rate has an important role in the distribution of the products of cellulose pyrolysis. The application of the low heating rates favors the dehydration to the bio-char, while the high heating rates favor the depolymerization to the levoglucosan. The first part of above-indicated fact usually occurs at temperatures below 350°C, while the second part that includes depolymerization leading to anhydro-sugars (levoglucosan) occurs in the temperature range of 300 – 450°C, as an integrated mechanism in "active" cellulose degradation. All of these are in good agreement with Waterloo mechanism of cellulose decomposition (Van de Velden et al. 2010).

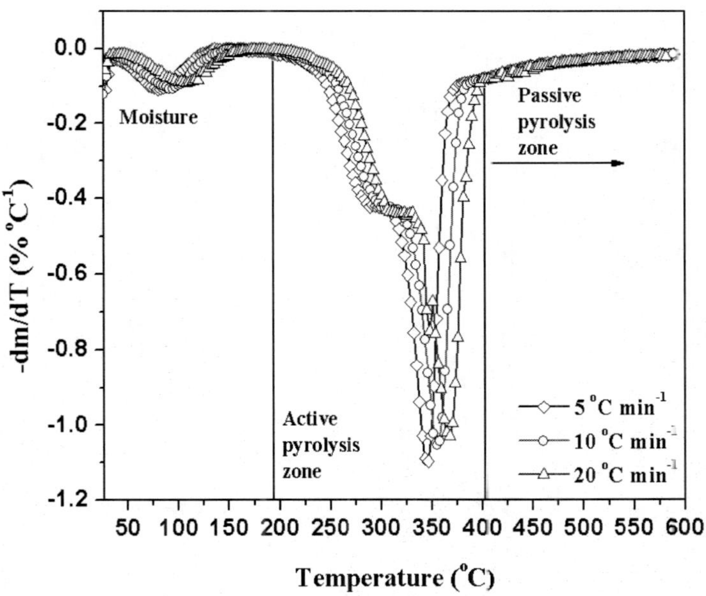

Figure 2. DTG (derivative thermogravimetry) curves of pyrolysis process of MDF derived Beech sawdust, at the various heating rates (5, 10 and 20oC min-1) in a nitrogen atmosphere.

The final stage involves lignin decomposition at temperatures above 450°C denoted by the long tailing in DTG curves (Figure 2). It can be observed that an increase in heating rate shifted DTG (also and TGA) curves to higher temperatures. This effect is known as the thermal hysteresis or thermal lag. That was because higher heat transfer efficiency would be achieved at the lower heating rate. In addition to actual considerations, the adhesive UF decomposition may be expected in an active pyrolysis zone, where the additional very small peak was observed at DTG curve at the highest heating rate (20°C min^{-1}) at about 347.9°C (Figure 2). This statement is in accordance with the thermal stability of UF resin (Ahamad and Alshehri 2014).

In this context, the presence of UF adhesive may lead to increased content of volatile matter (VM), since the higher yields of VM can be attributed to pyrolysis reactions of unstable structures of UF adhesive. Namely, as interesting characteristic, it should be mentioned that presence of adhesive probably affects the reduction of the maximal mass loss rate with an increase of heating rate, which is related to cellulose decomposition (Figure 2).

Therefore, the presence of UF adhesive in MDF derived Beech sawdust affects the "slows down" the decomposition of cellulose in an active zone, and this is probably reflected in a suppression of the rate of pyrolysis. Creation of an increased alkaline reaction environment caused by UF adhesive may be an important reason for "mitigation" of the cellulose decomposition. From obtained results, the presence of UF adhesive has no significant impact on hemicelluloses decomposition behavior (Figure 2). On the other hand, the influence of UF adhesive on the lignin decomposition can be reflected through the slightly increasing of the maximum mass loss rate, but with decreasing in the final bio-char yield (Zhang et al. 2016). However, the "degree" of the influence of the UF adhesive on thermal and kinetic characteristics of the pyrolysis process of MDF derived Beech sawdust depends on its percentage representation in the raw material.

Evaluation of the Impact of Process Parameters (Heating Rate, Temperature and Time) on TGA Pyrolysis Profiles through Matrix Analysis

Figure 3 (a), (b) and (c) shows the 3D representation of Mass — Temperature — Time dependence with the spatial matrix network, estimating through an alternating mathematical order with variables such as temperature (T) and time (indicated by τ). The mathematical order coefficients were elements of the formed matrix of the process at every considered heating rate. Details of the general stated model and algorithm are given below:

Code 1 - Algorithm for evaluation of 3D matrix representation of Mass-Temperature-Time dependency expressed through alternating mathematical order at the heating rate of $\beta = 5°C\ min^{-1}$.

Input file loaded: Heating rate = 5°C min^{-1}

General model:

$$Mass = p_{00} + p_{10}\cdot T + p_{01}\cdot\tau + p_{20}\cdot T^2 + p_{11}\cdot T\cdot\tau + p_{30}\cdot T^3 + p_{21}\cdot T^2\cdot\tau + p_{40}\cdot T^4 + p_{31}\cdot T^3\cdot\tau + p_{50}\cdot T^5 + p_{41}\cdot T^4\cdot\tau + p_{60}\cdot T^6 + p_{51}\cdot T^5\cdot\tau + p_{70}\cdot T^7 + p_{61}\cdot T^6\cdot\tau + p_{80}\cdot T^8 + p_{71}\cdot T^7\cdot\tau + p_{90}\cdot T^9 + p_{81}\cdot T^8\cdot\tau + p_{100}\cdot T^{10}$$

where T is normalized by mean 323.1 and std 157.4 and τ - time is normalized by mean 59.89 and std 30.56.

Coefficients (with 95% confidence bounds):

p00 = 149.5 (137.3, 161.7)
p01 = 4463 (3839, 5087)
p10 = -4374 (-4990, -3758)
p100 = 26.85 (18.79, 34.91)
p11 = 9789 (8490, 1.109e+04)
p20 = -1.003e+04 (-1.133e+04, -8742)
p21 = -5190 (-9136, -1243)
p30 = 4728 (792.4, 8664)
p31 = -2.209e+04 (-2.457e+04, -1.962e+04)

p40 = 2.22e+04 (1.968e+04, 2.472e+04)
p41 = -8075 (-1.117e+04, -4984)
p50 = 8638 (5553, 1.172e+04)
p51 = 1.415e+04 (1.158e+04, 1.671e+04)
p60 = -1.399e+04 (-1.658e+04, -1.14e+04)
p61 = 8704 (6478, 1.093e+04)
p70 = -8972 (-1.122e+04, -6723)
p71 = -2703 (-3449, -1956)
p80 = 2567 (1831, 3303)
p81 = -1909 (-2496, -1323)
p90 = 1953 (1359, 2547)

gof = struct with fields:
sse: 99.1295
rsquare: 0.9995
dfe: 173
adjrsquare: 0.9994
rmse: 0.7570

$$M = \begin{bmatrix} 149.48 & 4463.18 & -4374.27 & 26.85 \\ 9789.50 & -10033.46 & -5189.50 & 4728.10 \\ -22094.68 & 22203.96 & -8075.31 & 8637.83 \\ 14147.86 & -13990.85 & 8703.76 & -8972.33 \\ -2702.63 & 2566.98 & -1909.50 & 1953.08 \end{bmatrix}_{5x4}$$

$m = 5$ (rows)
$n = 4$ (columns)
Skew-Asymmetric matrix

Code 2 - Algorithm for evaluation of 3D matrix representation of Mass-Temperature-Time dependency expressed through alternating mathematical order at the heating rate of $\beta = 10^{o}C\ min^{-1}$.

Input file loaded: Heating rate = $10^{o}C\ min^{-1}$
General model:

$$Mass = p_{00} + p_{10} \cdot T + p_{01} \cdot \tau + p_{20} \cdot T^2 + p_{11} \cdot T \cdot \tau + p_{30} \cdot T^3 + p_{21} \cdot T^2 \cdot \tau + p_{40} \cdot T^4 + p_{31} \cdot T^3 \cdot \tau + p_{50} \cdot T^5 + p_{41} \cdot T^4 \cdot \tau + p_{60} \cdot T^6 + p_{51} \cdot T^5 \cdot \tau + p_{70} \cdot T^7 + p_{61} \cdot T^6 \cdot \tau + p_{80} \cdot T^8 + p_{71} \cdot T^7 \cdot \tau + p_{90} \cdot T^9 + p_{81} \cdot T^8 \cdot \tau + p_{100} \cdot T^{10}$$

where T is normalized by mean 319.7 and std 156.7 and τ - time is normalized by mean 30.68 and std 14.82.

Coefficients (with 95% confidence bounds):

p00 = 103.9 (92.58, 115.2)
p01 = 1149 (749.7, 1548)
p10 = -1143 (-1531, -755.3)
p100 = 33.88 (22.01, 45.76)
p11 = 3373 (2367, 4380)
p20 = -3325 (-4332, -2318)
p21 = 7583 (5337, 9829)
p30 = -7820 (-1.003e+04, -5611)
p31 = -1.118e+04 (-1.445e+04, -7897)
p40 = 1.072e+04 (7455, 1.398e+04)
p41 = -1.836e+04 (-2.252e+04, -1.421e+04)
p50 = 1.868e+04 (1.445e+04, 2.292e+04)
p51 = 5987 (3223, 8752)
p60 = -5400 (-8197, -2604)
p61 = 1.01e+04 (7372, 1.282e+04)
p70 = -1.022e+04 (-1.297e+04, -7475)
p71 = -616.3 (-1835, 602.1)
p80 = 360.6 (-865.4, 1587)
p81 = -1450 (-2085, -815.7)
p90 = 1461 (811.2, 2112)

gof = struct with fields:
sse: 138.1050
rsquare: 0.9993
dfe: 173
adjrsquare: 0.9992
rmse: 0.8935

$$M = \begin{bmatrix} 103.87 & 1148.86 & -1143.07 & 33.88 \\ 3373.27 & -3324.70 & 7583.10 & -7820.27 \\ -11175.88 & 10715.26 & -18362.78 & 18683.58 \\ 5987.45 & -5400.33 & 10096.45 & -10224.20 \\ -616.32 & 360.64 & -1450.18 & 1461.50 \end{bmatrix}_{5x4}$$

$m = 5$ (rows)

$n = 4$ (columns)

Skew-Asymmetric matrix

Code 3 - Algorithm for evaluation of 3D matrix representation of Mass-Temperature-Time dependency expressed through alternating mathematical order at the heating rate of $\beta = 20^{\circ}$C min^{-1}.

Input file loaded: Heating rate = 20°C min^{-1}

General model:

$$Mass = p_{00} + p_{10} \cdot T + p_{01} \cdot \tau + p_{20} \cdot T^2 + p_{11} \cdot T \cdot \tau + p_{30} \cdot T^3 + p_{21} \cdot T^2 \cdot \tau + p_{40} \cdot T^4 + p_{31} \cdot T^3 \cdot \tau + p_{50} \cdot T^5 + p_{41} \cdot T^4 \cdot \tau + p_{60} \cdot T^6 + p_{51} \cdot T^5 \cdot \tau + p_{70} \cdot T^7 + p_{61} \cdot T^6 \cdot \tau + p_{80} \cdot T^8 + p_{71} \cdot T^7 \cdot \tau + p_{90} \cdot T^9 + p_{81} \cdot T^8 \cdot \tau + p_{100} \cdot T^{10}$$

where T is normalized by mean 315.5 and std 153.6 and τ - time is normalized by mean 15.92 and std 6.87.

Coefficients (with 95% confidence bounds):

p00 = 106.8 (84.01, 129.7)
p01 = 691 (130.1, 1252)
p10 = -725 (-1210, -239.7)
p100 = 33.69 (20.01, 47.36)
p11 = 215 (-1600, 2030)
p20 = -214.9 (-1949, 1519)
p21 = 5201 (2683, 7720)
p30 = -5373 (-8001, -2744)
p31 = -7001 (-9121, -4882)

p40 = 6712 (4766, 8658)
p41 = -8892 (-1.281e+04, -4969)
p50 = 9294 (5306, 1.328e+04)
p51 = 6676 (4645, 8706)
p60 = -6284 (-8328, -4240)
p61 = 5606 (3613, 7598)
p70 = -5860 (-7862, -3859)
p71 = -1379 (-2129, -628.7)
p80 = 1176 (388.1, 1963)
p81 = -1030 (-1342, -718.5)
p90 = 1075 (750.9, 1399)

gof = struct with fields:
sse: 171.2710
rsquare: 0.9991
dfe: 172
adjrsquare: 0.9990
rmse: 0.9979

$$M = \begin{bmatrix} 106.85 & 690.99 & -724.97 & 33.69 \\ 215.02 & -214.86 & 5201.34 & -5372.69 \\ -7001.33 & 6711.97 & -8892.00 & 9293.59 \\ 6675.59 & -6283.64 & 5605.72 & -5860.49 \\ -1378.97 & 1175.65 & -1030.18 & 1074.90 \end{bmatrix}_{5x4}$$

$m = 5$ (rows)
$n = 4$ (columns)
Skew-Asymmetric matrix

By changing the dimension of the matrix of the type "$m \times n$", the influence of process parameters on the independent-variable (mass of the sample, m), which essentially forms the thermo-analytical curve, as well as the extent of its displacement in the 3D space, can be estimated.

From obtained results, in all cases, the skew-asymmetric matrix with dimensions (5 × 4) was obtained, so the shape of all TGA curves at all heating rates is the same. However, the part of the row vectors whose elements are coefficients of alternating mathematical orders at various heating rates, differ from each other in a sign. The coefficients have different values but differ in the sign. Namely, the signs of individual coefficients in the parts of row vectors are the same type ("+" or "–") and in the same positions, at the heating rates of 10 and 20°C min^{-1}, as **B** (10°C min^{-1}) = (+7583.10 –7820.27), **C** (20°C min^{-1}) = (+5201.34 –5372.69), but for 5°C min^{-1}, the signs are not in the same position, they are replaced "seats", as **A** (5°C min^{-1}) = (–5189.50 +4728.10). This difference, in respect to present magnitude values (coefficient values) and 'positions' of the same type of sign, dictate the extent of the surface of matrix network shown in Figure 3 (a), (b), and (c).

Absolute values of the coefficients in parts of row vectors related to heating rates of 10 and 20°C min^{-1}, are higher than those related to a heating rate of 5°C min^{-1}, and this leads to narrowing the surface of matrix network.

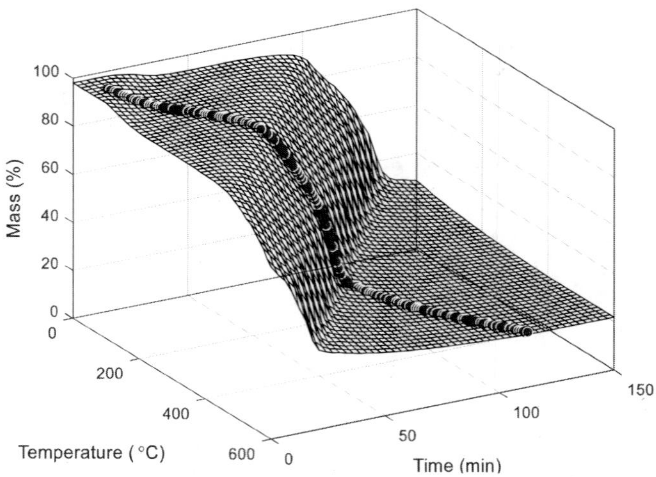

Figure 3. (a) The 3D representation of mass — temperature — time dependence with the spatial matrix networks at 5oC min-1.

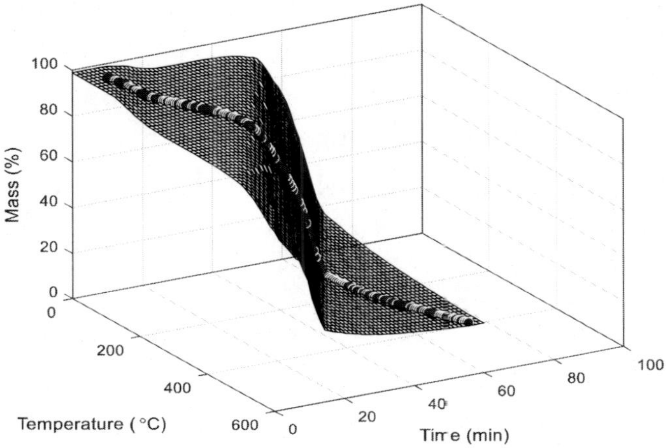

Figure 3. (b) The 3D representation of mass — temperature — time dependence with the spatial matrix networks at 10oC min-1.

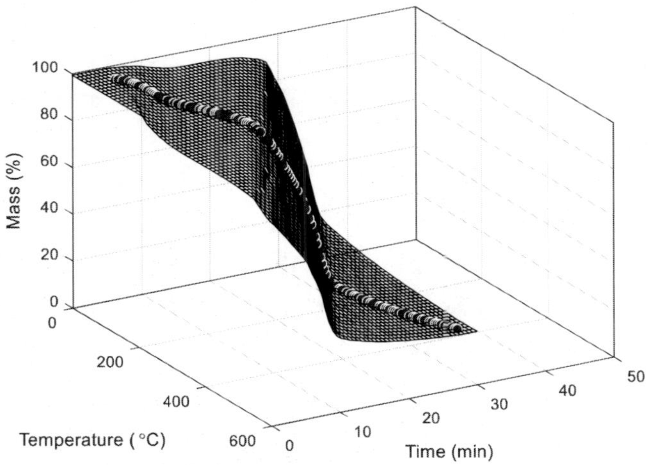

Figure 3. (c) The 3D representation of mass — temperature — time dependence with the spatial matrix networks at 20oC min-1.

For heating rates of 10 and 20°C min^{-1}, the degree of narrowing is approximately the same (Figure 3 (b) and (c)), and this narrowing indicates that the medium and higher heating rates introduce a greater uncertainty in the sense that the actual heating rates (and appropriate temperatures at these heating rates) are not desirable for MDF derived Beech sawdust pyrolysis, and thus the decomposition of the lignocellulosic material.

Changing the position of the signs at the lowest heating rate is crucial in the rapid expansion of the surface of matrix network (Figure 3 (a)), which rapidly increases the reliability of performing the quality pyrolysis process of the studied system, precisely at a given heating rate. So, for the optimal pyrolysis process of the MDF derived Beech sawdust very low heating rates are highly desirable, but with a longer residence time. Namely, at high heating rates of the TGA measurements, the true temperature of the tested sample can be distorted, in the sense that it may become unknown.

The temperature stability linked with the studied sample is much more precise at low heating rates for TGA testing. For the high heating rates and much larger particle sizes, the monitored process may be no longer kinetically controlled and thermal gradients can be substantially expressed. For lower heating rates (< 10°C min^{-1}), it can be observed occurrences the elevation of the effectiveness of the heat transfer to the particles. Therefore, reaching the higher temperatures are needed to provide particles the minimum heat required, so that their cracking process starts, which provokes the certain delay in the process in terms of operating temperature. In respect to the actual decomposition reactions which would be done, with the application of low heating rates, the possibility of the secondary reactions which take place can be significantly reduced. In this way, the use of low heating rates would ensure that “thermal cracking” does not occur, so we expect a high yield of bio-char. In contrast, the use of high heating rates will then favor the production of liquid and gas products over the appearance of fragmentation reactions of tested biomass, giving at the same time a limited yield of bio-char. These boundary changes can be clearly seen in Figure 3 (a), (b) and (c) for our tested reaction system, using a matrix analysis. However, in the context of kinetic analysis, it should be pointed out that if the decomposition process was governed by the kinetically controlled regime, the actual process itself is independent on the heating rate, the particle size, or the experimental temperature. So, the reactions take place under kinetic regime conditions when the reaction rate is only dependent on and determined by the intrinsic kinetics of particles. This phenomenon occurred when the heat transfer to the particles composing the sample and the heat transfer within each of the particles is the fast in the comparison to the decomposition (process) rate (Willner and Brunner 2005).

In such cases, it can usually be unsubstantial the heat transfer limitations, and at the same time, the mass transfer limitations can be assumed to be negligible, if the mentioned conditions for the kinetic modeling are suppositional.

Gaseous Products Analysis

Study of the evolution of the gases emitted during pyrolysis of tested material is realized by analyzing the changes of ionic current obtained by mass spectrometry in a function of experimental temperatures. The used mass spectrometry (MS) analysis for the pyrolysis process of the investigated biomass material that generates the volatile products is applied online through direct mass spectrometry scanning.

Thermogravimetric Analysis

The mass spectrometry (TGA-MS) profiles of the evolved gases during the pyrolysis process, such as H_2, CH_4, C_2H_6, C_4H_{10}, C_5H_{12} and profiles of the changes in the ion current as a function of temperature (the mass spectra analysis) for the pyrolytic water (H_2O), CO_2 (C_3H_8) and CO for the MDF (the medium-density fiberboard) derived Beech sawdust samples are shown in Figure 4 a), b), c), d), e) and f).

The present study was focused on the main volatile products of the tested biomass pyrolysis, based on the changes in the ion currents across the experimental temperature range of ΔT = 40 — 600°C, and on their relevancy, assigned to the ion/mass intensities (*m/z*) with corresponded screening analysis performed in the selected-ion monitoring (SIM) mode (the ions characteristic of each molecules, such as follows: 2, 16, (17), 18, 28, 44, ... for the H_2, CH_4, (NH_3), H_2O, CO, CO_2, ... were monitored) and which is in the accordance with the database of the National Institute of Standards and Technology (in abbreviated form: NIST).

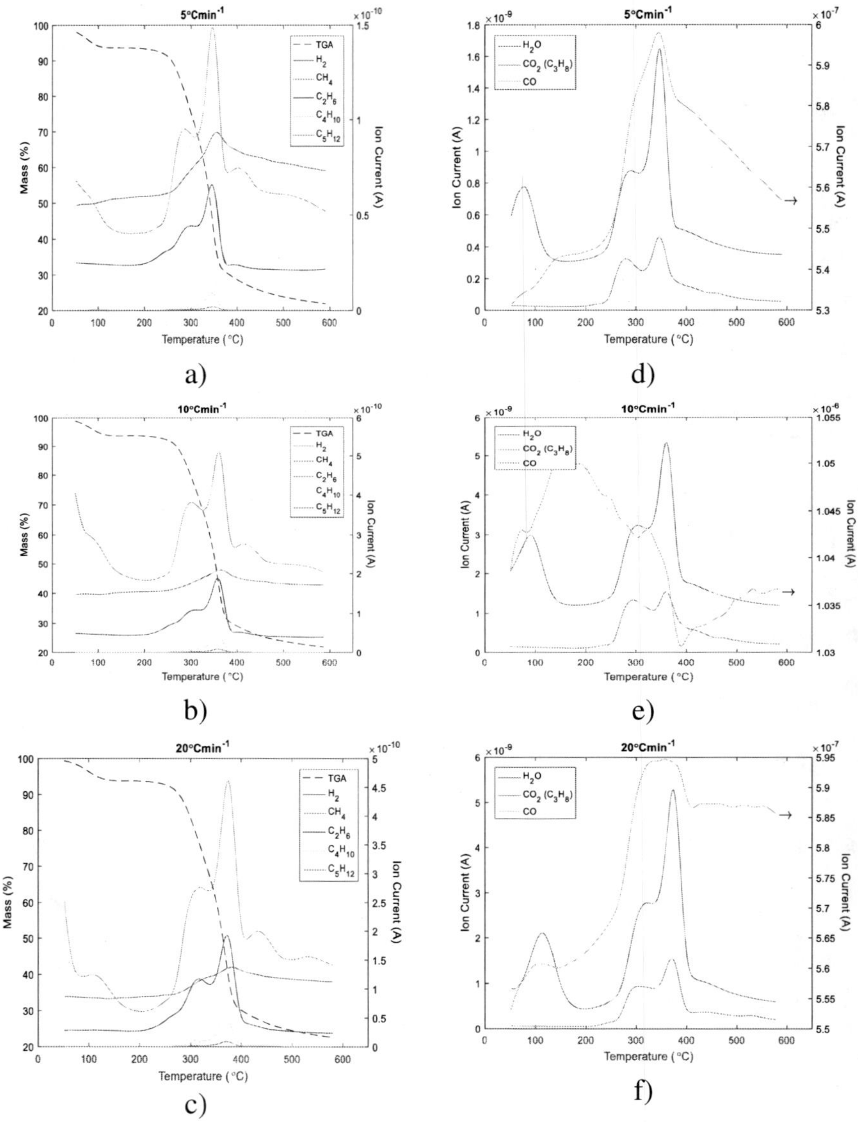

Figure 4. TGA-MS profiles of evolved gases during pyrolysis process (H2, CH4, C2H6, C4H10, C5H12, H2O, CO2 (C3H8) and CO) presented as changes in ion current as a function of temperature for MDF derived Beech sawdust samples.

From all presented results, the eventual emission of the NH_3 as a product of the nitrogen transformation that arises from the UF adhesive compound was not detected, due to its very low concentration present in the sample, so that during the analysis by the mass spectrometry this value was below the detection limit, and therefore not shown.

Also, having in mind that the nitrogen content in the investigated sample is extremely small (almost negligible), therefore it was below the detection limit for the current analysis. Therefore, the presented MS spectra show the evolved gas analysis of all above-indicated gases realized during the actual process.

From obtained results (Figure 4), including all heating rates, the devolatilization stage was identified with major mass loss and the release of organic compounds which leads to the formation of bio-char, where this stage releases H_2, CH_4, H_2O, CO_2 (also C_3H_8), CO, and some light hydrocarbons. Namely, the "maximums" profiles in DTG curves (Figure 2) approximately follow the maximum gas releases, as shown in Figure 4. However, the only drastic departure from this behavior occurs in the case of CO evolution at all heating rates (Figure 4 d), e) and f)). Generally, with an increase of the heating rate, the 'intensity' of the ion current also increases, so that the rate of release of the gas products is higher. Unlike the other gas species, it can be observed that the evolution of CO gas drastically changes with the change in the heating rate, where taking into account wide temperature range of lignin decomposition, the core of CO releases may originate from cellulose and lignin degradation (Figure 4 d), e) and f)). So, except for the CO, the gas evolution profiles (Figure 4) are relatively symmetrical ones and follow DTG profiles. Bearing in mind the above–discussion, the lignin has a significant role and its contribution largely depends on the rate of heating. Since the lignin is decomposed continuously, with the un-defined rate of the mass loss, such behavior is manifested through a series of "unsteady" profiles of CO evolution (Figure 4 d), e) and f)). It should be noted that the CH_4 and H_2 peaks are detected in a passive pyrolysis zone for $T > 400°C$, and such secondary pyrolysis is the result of decomposition of heavy molecules in bio-char. The H_2

evolved is, in fact, only a part of hydrogen present due to the higher volatile matter content in the sample.

However, if we severely monitor the DTG and MS features, contrary to the CO_2 and water profiles, CH_4 and H_2 profiles (Figure 4 a), b) and c)) do not strictly follow the DTG profiles, while their evolution starts at greater temperatures. With regard to H_2, the charring and restructuring processes, which include the dehydrogenation reactions, are responsible for its production, starting at the temperature above 450°C. The relatively large amounts of H_2O, CO_2 and CO (Figure 4 d), e) and f)) are formed because of a large number of hydroxyl groups and oxygen atoms present in considered biopolymers that make up the cell walls (cellulose, hemicelluloses, and lignin).

The H_2O profile occurs over the entire temperature range of the sample pyrolysis, and this indicates on the multiple origins of the water. The release of adsorbed water related to the peak around 100°C, whereas the main peak occurs above 300°C, which can be attributed to the water released primarily from the cellulose decomposition. In the temperature region of 290 – 400°C, the CO_2 and CO release also can be formed from cellulose and hemicelluloses decomposition reactions. The peak around 350°C for both, CO_2 and CO, can be ascribed to the decomposition of aromatic and aliphatic carboxyl groups in cellulose.

The CO evolution profile above 400°C can be attributed to the bio-char formation reactions, in which the residual oxygen–functional groups are cleaved. In addition, as the temperature increased, the more stable ether structures and oxygen-bearing hetero–cycles in lignin are also decomposed into CO_2. The CH_4 evolution profile exhibits two main features at around 350 and 420/430°C (Figure 4 a), b) and c)) which correspond to governed processes of CH_4 production. The first one probably can be attached to decomposition of lignin methoxyl groups (Jakab et al. 1997), while the CH_4 evolution at the higher temperature can be attributed to charring processes (involve aliphatic and aromatic chars). It can be noted that charring reactions for the tested system does not play such a significant role and this is can be confirmed with not so pronounced the H_2 evolution (especially at the highest heating rate) (Figure 4 a), b) and c)).

Compared to the hemicelluloses and lignin, cellulose released most of the gaseous products in a narrow temperature range (Figure 4). Since that CH_4 reaches the maximum amount around 380°C, the CH_4 emission at lower temperatures was most likely released from C—C bond cleavage in aliphatic chains, whereas CH_4 emission at higher temperatures was mainly produced from cracking of a weakly bonded methoxyl—O—CH_3 group, as well as the break of having higher bond energy of methylene group —CH_2 (Ferdous et al. 2001).

The H_2 formation which arises at high temperatures (T > 400°C), can be attributed to secondary reactions as the steam reforming and/or tar cracking reactions (Widyawati et al. 2011). At the lower temperatures, the H_2 emission was mainly generated from the decompositions of cellulose and hemicelluloses. At T > 400°C, the decreasing trends in the H_2 mass signal may reveal that lignin pyrolysis could also generate the hydrogen.

Isoconversional ("Model-Free") Analysis

Figure 5 shows the dependence of effective activation energies ($E_{a,\alpha}$) on the conversion fraction (α) with indicated 95% confidence interval limits for pyrolysis process of MDF derived Beech sawdust, using the Vyazovkin's advanced isoconversional method. In Figure 5, the variation of $E_{a,\alpha}$ on α is shown, excluding the part related to the water evaporation.

The pyrolysis process of MDF derived Beech sawdust exhibit a significant variation of $E_{a,\alpha}$ on α, where three specific reaction zones were identified (Figure 5). In the first reaction zone ("1"), the $E_{a,\alpha}$ varies with increasing trend from 155.8 (α ~ 0.15) up to 188.4 kJ mol^{-1} (α ~ 0.25), and further with a relatively stable value up to α ≈ 0.33, enclosing the mean value of $E_{a,\text{mean}(1)}$ = 179.7 kJ mol^{-1}. The second reaction zone ("2") includes the initial increase of $E_{a,\alpha}$ value from 189.1 (α ~ 0.34) up to 221.9 kJ mol^{-1} (α ~ 0.47), then the $E_{a,\alpha}$ value decreases to 196.9 kJ mol^{-1} (α ~ 0.63), with a mean value of $E_{a,\text{mean}(2)}$ = 207.5 kJ mol^{-1} (Figure 5). Finally, the third reaction zone ("3") is characterized by the trend of increasing $E_{a,\alpha}$ values from $\alpha \sim 0.64$ (197.3 kJ mol^{-1}) up to the very end of the process at $\alpha = 0.85$ (322.3 kJ mol^{-1}), with a mean value of $E_{a,\text{mean}(3)} = 223.2$ kJ mol^{-1}.

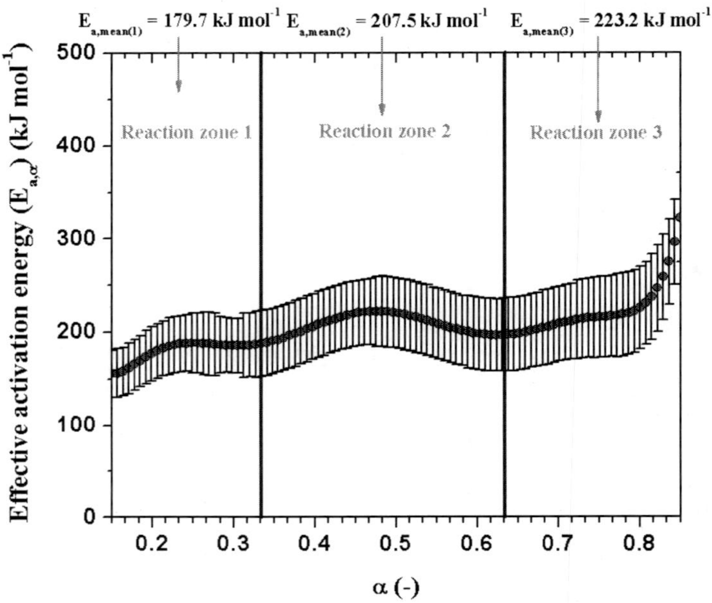

Figure 5. Dependence of effective activation energies ($E_{a,\alpha}$) on the conversion fraction (α) with indicated 95% confidence interval limits, for pyrolysis process of MDF derived Beech sawdust, using the advanced isoconversional method.

The first reaction zone encompasses the temperature region of $\Delta T_{(1)}$ = 263 – 321°C typically for hemicelluloses depolymerization. The actual increase in the $E_{a,\alpha}$ value in the conversion range of $\alpha = 0.15 - 0.25$ can be attributed to hemicelluloses decomposition (55 – 187 kJ mol^{-1} (Grønli et al. 2002)), within temperature interval of 263 – 295°C, corresponded to the "shoulder" of the DTG curve (Figure 2). The relatively stable range of $E_{a,\alpha}$ in a narrow conversion range ($\alpha = 0.25 - 0.33$ within $\Delta T_{(1)}$ = 295 – 321°C) belongs to slowly transition from exiting from the first and entering into another, the second reaction zone, corresponding to the transition of the "shoulders" into "well-defined" peaks on the DTG curves, shown in Figure 2. This can be attached to the beginning of cellulose decomposition, but further progress of the process is followed by a stronger variation of the effective activation energy on the conversion, within the second reaction zone (Figure 5). The obtained mean value of 207.5 kJ mol^{-1} (Figure 5) is close to the value of E_a that corresponds the decomposition of the pure crystalline cellulose (238 kJ mol^{-1}) (actual range 195 – 238 kJ mol^{-1}) (Mamleev et al. 2006), but with assumed the first-order mechanism (Antal and Várhegyi 1995). However, the cellulose decomposition in actual reaction system within second reaction zone is the nearest to the situation that corresponds to the multi-step decomposition process, in accordance with validation through the Broido-Shafizadeh model (Kilzer and Broido 1965; Bradbury et al. 1979). Namely, at the beginning, the increasing trend of E_a for cellulose decomposition (Figure 5) ($\Delta T_{(2)}$ = 323 – 340°C) in a lower-temperature region, probably includes the reduction in degree of polymerization by bond-scission, the occurring of free radicals with water elimination (Figure 4 d), e) and f)), formation of gases with low molecular weight (such as CO and CO_2) and subsequently obtaining the char residues. The second part, where the E_a values are reduced ($\Delta T_{(2)}$ = 343 – 361°C), may involves competitive pathways, where primarily reaction in the mechanism can include depolymerization by trans-glycosylation, followed by the dehydration and formation of the char and gases (Bradbury et al. 1979). Subsequent secondary reactions most often describe the degradation of tar to lighter gases and/or condensation to the solid material (Antal 1983).

It can be pointed out that the estimated mean value related to the second reaction zone ($E_{a,\text{mean}(2)}$ = 207.5 kJ mol^{-1}, Figure 5) is very close to that value which corresponds to depolymerization in Broido-Shafizadeh model (198 kJ mol^{-1}) and this value is higher from the ones envisaged for dehydration (153 kJ mol^{-1}) (Bradbury et al. 1979). Thus, a lower temperature favors the dehydration creating primarily char and CO_2, while the depolymerization reaction, probably will be favored at the higher heating rates with an accelerating character and in a reduction of E_a's, in the first place, the liberation of the gas.

The third reaction zone ("3") which is accompanied by an increase in the $E_{a,\alpha}$ value (Figure 5) can be related to lignin decomposition. Since that lignin is mainly composed of three kinds of benzene-propane which were heavily cross-linked, these facts make it have a very high thermal stability.

Therefore, at a later stage of pyrolysis, in the higher temperature region, the part of three-dimensional network structure broke, which requires a much higher E_a values. The obtained mean value of E_a related to actual reaction zone ($E_{a,\text{ mean(3)}} = 223.2$ kJ mol^{-1}) (actual range 35 – 267 kJ mol^{-1} (Grønli et al. 2002)) is much higher than E_a value for the pure alkali lignin (70.7 kJ mol $^{-1}$) (Pasangulapati et al. 2012). Taking this into account, the decomposition of the lignin takes place much slower than the hemicelluloses and cellulose decomposition, even if it covers a wider temperature range of decomposition (~ 160 – 900°C (th) ("th" – theoretically substantial)). From this issue, it follows that much higher E_a value obtained for third reaction zone and related here for lignin decomposition can be attributed to the great variety of the bonds and multi-phasic character of its conversion. When the high temperature is reached, due to the increasing of aromatic character of lignin derived char, which raises its thermal stability, the lignin is transformed into the highly cross-linked carbonaceous material (Figure 5). In addition, the lignin obtained from the Klason method exhibits the higher condensed structure (Klein and Virk 2008), when compared with lignin obtained from other separation methods, which additionally confirms the above-mentioned facts.

Furthermore, the apparent activation energy of the UF resin decomposition falls within the range of $\Delta E_a = 150 - 300$ kJ mol^{-1} (Zhang et al. 2016) but in highly heterogeneous lignocellulosic system as studied, it can be expected that these value can be contributed eventually on the considered cellulose decomposition. So, the UF adhesives can affect the first part of this process (in a lower temperature zone) promotes the chemical bonds fracture in cellulose and accelerates the formation of radicals (such as carbonyl or hydroxyl radicals). Taking into account the above-estimated values of E_a and actual values for the UF resins, the adhesive may also influenced on the lignin decomposition through eventual promotion of releasing the gases during pyrolysis, such as CO (Figure 4 d), e) and f)). However, these were assumptions, but based on the "model-free" analysis, but much more detailed analysis is needed, which will be presented in the following sub-sections.

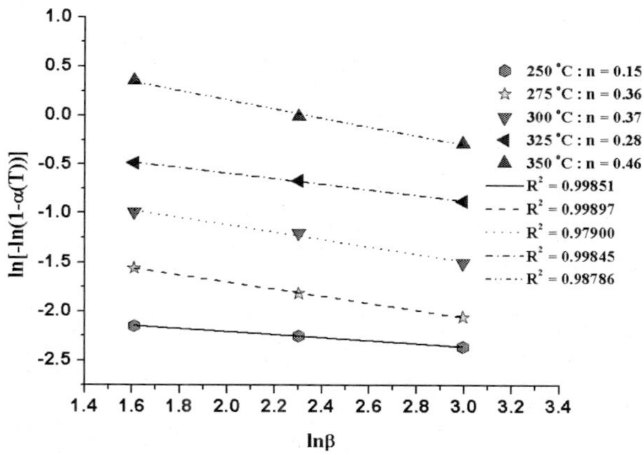

Figure 6. The ln[-ln(1-α(*T*))] *versus* ln*β* plots, for pyrolysis process of MDF derived Beech sawdust.

Avrami's Double-Logarithmic Plot Analysis

Figure 6 shows the ln[-ln(1-α(*T*))] against ln*β* plots, for the purpose of comparison of the Flynn-Wall-Ozawa (FWO) exponents. It was assumed that the Flynn-Wall-Ozawa exponent depends on the temperature.

It can be observed from Figure 6 that FWO values were close to the zero, and values are scaled in the range of n = 0.15 – 0.46. Also, the significant variation of n with temperature was identified. Likewise, it can be noted that FWO values were dependent on the conversion fraction, and FWO exponent is not constant during the pyrolysis process. Such a phenomenon is a consequence of the multiple reaction steps for MDF derived Beech sawdust pyrolysis. It can be seen from Figure 6 that with gradually increasing of temperature, the n value also increases (except at a temperature of 325°C). This behavior and a sudden jump of n value from 0.28 to 0.46 with increased trend of temperature, may suggests that for studied pyrolyzed system, the initial temperature for cellulose decomposition moves towards a higher temperature region, since that the high amount of cellulose is present in the sample (Table 1). It should be noted that the biomass system which possess the high percentage of ash and impurities is distinguished by a lower initial temperature of the decomposition, and this behavior is not a feature of our examined system (Table 1). Switching to the high temperature values introduces high n values, which is followed by the lignin decomposition (see also Figure 5). Generally, the trend of changing the FWO values with the temperature monitors the E_a changes with the conversion fraction.

Simulation of Pyrolysis Process Assuming Kinetically Controlled Regime without Influence of UF Adhesive

The pyrolysis process of MDF derived Beech sawdust was numerically simulated including three-step kinetics model, excluding the adhesive wood component. The simulation was performed under the assumption of the existence of a kinetically controlled process without interfering the diffusion effects (when $E_a^{\mathrm{Diff}} \to 0$), using the kinetic parameters ($E_{a,\,\mathrm{mean}}$) derived from isoconversional analysis. The simulation was carried out on the basis of the equation:

$$-\ln\left(1-\alpha_{Kin}\right)=\left(\frac{A_{Kin}RT_{\beta}^{2}}{\beta E_{a,mean}}\right)\exp\left(-\frac{E_{a,mean}}{RT_{\beta}}\right), \tag{16}$$

and after a certain rearrangements of the upper equation, finally, the expression for the theoretical conversion fraction value under kinetically controlled regime can be obtained as:

$$\alpha_{Kin}=1-\exp\left[-\frac{A_{Kin}\cdot R\cdot T_{\beta}^{2}}{\beta\cdot E_{a,mean}}\exp\left(-\frac{E_{a,mean}}{RT_{\beta}}\right)\right], \tag{17}$$

where A_{Kin} is the pre-exponential factor [min^{-1}] assuming kinetically controlled process, T_β is the value of the experimental temperature at a considered heating rate, β, while $E_{a,\mathrm{mean}}$ is the mean effective activation energy obtained from the isoconversional analysis. The simulation was performed using the separate conversion fraction sectors for pseudo-components (hemicelluloses, cellulose, and lignin) decomposition processes, without the impact of the UF resin adhesive in Beech sawdust sample. Table 2 lists the values of kinetic parameters obtained through the simulation procedure using the Eq. (17), with strictly indicated conversion fraction values obtained by isoconversional approach, at every considered heating rate ($\beta = 5$, 10 and 20°C min^{-1}).

The obtained values and their magnitudes of A_{Kin} associated with corresponding pseudo-components are in the range of the pre-exponential factors typical for the complex, multi-component mechanism of pyrolysis process attached to the lignocellulosic material, such as wood/woody biomass (Di Blasi 2008; Wang et al. 2008). However, it should be noted that the kinetic parameters of lignin decomposition (Table 2) strongly depends on the extraction method from lignocellulosic material. The heat and mass transfer processes can significantly affect on effective activation energy and pre-exponential factor for lignin decomposition. Effective activation energies in the range of $E_a \sim 129 — 361$ kJ mol^{-1} and 80 — 158 kJ mol^{-1} and pre-exponential factors in the range of $A \sim \times 10^{11} — \times 10^{22}$ min^{-1} and $\times 10^{7} — \times 10^{9}$ min^{-1} were reported (Brebu and Vasile 2010) for thermal decomposition of Alcell and Kraft lignins, respectively, at heating rates of 5 – 15°C min^{-1}.

Figure 7 a), b) and c) shows the experimental conversion fraction curves and simulated $\alpha_{Kin} — T$ plots in temperature intervals predicted for pseudo-components decomposition reactions, at every considered heating rate.

Table 2. The kinetic parameters for three-step kinetics model of MDF derived Beech sawdust pyrolysis without influence of UF-resin adhesive

β_i (°C min^{-1})	**i = 1, 2, 3 (1: 5°C min^{-1}; 2: 10°C min^{-1}; 3: 20°C min^{-1})**	**i = 1, 2, 3 (1: 5°C min^{-1}; 2: 10 °C min^{-1}; 3: 20°C min^{-1})**	**i = 1, 2, 3 (1: 5°C min^{-1}; 2: 10°C min^{-1}; 3: 20°C min^{-1})**
$\Delta\alpha$ [a]	I: 0.15 — 0.35	II: 0.35 — 0.65	III: 0.65 — 0.85
A_{Kin} (min^{-1})	1.50×10^{16}	2.80×10^{17}	3.20×10^{19}
$E_{a,mean}$ (kJ mol^{-1})	179.7	207.5	223.2

[a] The α (conversion fraction) sections were chosen approximately from isoconversional analysis.

Figure 7 clearly shows various stages of pyrolysis process (with the exclusion of the drying period - removal of moisture) related to decompositions of three main pseudo-components presented in studied sample, without influence of UF adhesive. The simulated α_{Kin} — T plots at all heating rates precisely show separated regions of decomposition reactions related to each pseudo-component (hemicelluloses, cellulose, and lignin), for kinetics area of $\alpha_{Kin} \in$ <0.14,1>. Figure 7 a), b) and c) also shows the specific temperatures situated as position of the "shoulder" on the DTG curve for every considered heating rate (see Figure 2), as well as exact positions and values of initial and final temperature for cellulose decomposition process. All the rest up to α_{Kin} = 1 belongs to lignin decomposition and bio-char formation. Both indicated temperatures related to cellulose pyrolysis, T_i and T_f, are shifted symmetrically and laterally with an increase in heating rate (Figure 7). All presented α_{Kin} — T plots very well describe the thermal properties of the experimentally obtained curves. Therefore, the parameters presented in Table 2 serve as good indicators for any preliminary analysis, but this refers to observing the decomposition of each pseudo-component individually, without considering their possible interaction, and contribution from the adhesive. From presented results, it is obvious that the reactivity of the "shoulder" and "peak" are differs (including also the last section related to lignin). It should also be noted that for pyrolysis of cellulose, the kinetic parameters presented in Table 2 (in particular, the mean value $E_{a,\text{mean}}$ = 207.5 kJ mol^{-1}), can be eventually attributed to the one of the pathways in a complex mechanism, such as the elimination reaction. This reaction pathway produces light gases with effective activation energy in the range of 190 – 210 kJ mol^{-1} (Mamleev et al. 2007) and this corresponds exactly to the temperatures in the range between T_i and T_f in Figure 7 (look for comparison and Figure 4 d), e) and f)).

From above facts, the effect of mutual interactions was not taken into account. It is assumed that during the pyrolysis process, the different components are decomposing independently, with no interactions. Therefore, the further investigation which would include the model substances (Hosoya et al. 2007; Gu and Catchmark 2013) is required in order to consider the mentioned interactions.

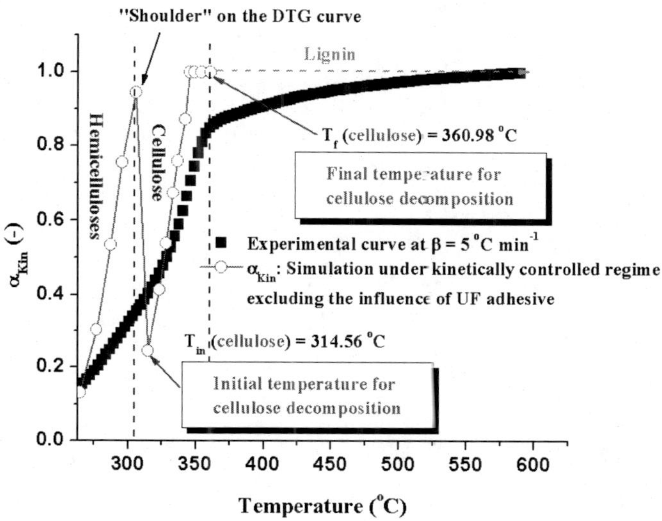

a)

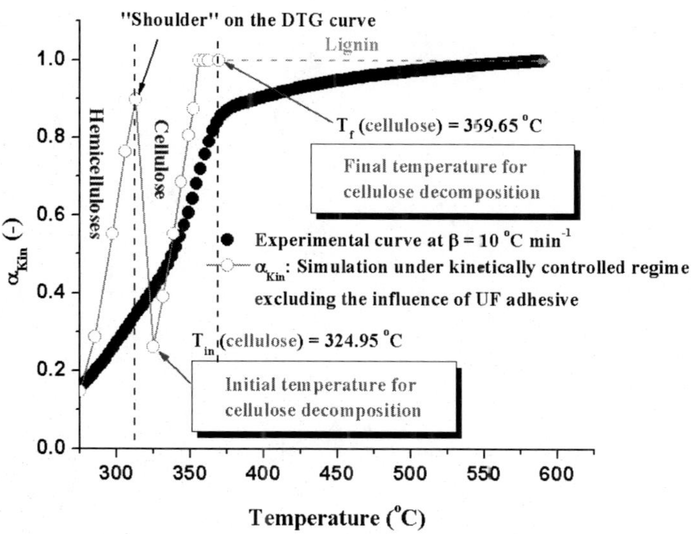

b)

Figure 7. (Continued).

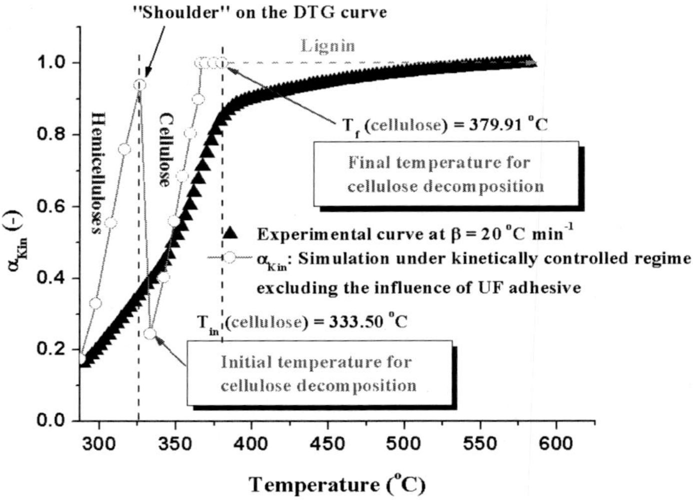

Figure 7. The experimental conversion fraction curves and simulated αKin — T plots in the temperature intervals predicted for pseudo-components decomposition reactions, at various heating rates.

CPRM Study for MDF Derived Beech Sawdust Pyrolysis

From established DTG curves (Figure 2) for pyrolysis process, it can be assumed that entire DTG profile of lignocellulosic material can be considered as overlaps of the pseudo-components DTG features. Such overlaps give a serious reason for the introduction of the parallel reaction model, which was proposed by Scheme 1. Taking this into account, it was carried out the optimization procedure assuming the existence of four components (hemicelluloses, cellulose, lignin, and UF resin wood adhesive) as composition chemical fractions of the tested biomass sample. In this approach, to explain the analysis result more clearly, hemicelluloses, cellulose, and lignin are considered as the main pseudo-components of studied lignocellulosic material, while UF resin adhesive was a fourth additional component. Therefore, in order to easily monitor the results, they are marked with: Co1 – Hemicelluloses, Co2 – UF-resin adhesive, Co3 – Cellulose, and Co4 – Lignin. The applied procedure includes an optimum numerical fitting accomplished by the non-linear regression that minimizes the difference between the measured and calculated data. The best optimization was performed through resolving the optimization problem which was given by the Eq. (10). The objective function has the following variables: the pre-exponential factor and effective activation energy, A_l and $E_{a,l}$ (l-th component), together with values of reaction orders n_l, making the entire set of kinetic parameters for each of the components; together with these optimized parameters, the model gives the best optimization value of the weighting factor, for each reaction in a parallel reaction mechanism.

Table 3 shows the CPRM parameters for the pyrolysis process of the MDF derived Beech sawdust sample. From established CPRM parameters, it can be observed that for all pseudo-components, the estimated kinetic parameters (A_l and $E_{a,l}$) are in good agreement with those derived from a preliminary analysis based on the "model-free" approach (Table 2), with pronounced contribution of the partial reaction attached to cellulose, but diminished contribution of the partial reaction attached to lignin (φ_l), considered all heating rates. However, a significant difference in kinetic parameters in relation to previously mention occurs in the case of UF-resin adhesive decomposition.

Namely, the kinetic parameters for the decomposition of the UF-resin adhesive are significantly different from those related to pseudo-components and are not in good agreement (Tables 2 and 3). However, it should be emphasized that the obtained kinetic parameters for the UF-resin adhesive decomposition (Table 3) during studied pyrolysis process are significantly lower than those predicted for decomposition of pure UF resin (Zhang et al. 2016). This can be a consequence of possible promoting effect in the decomposition reaction pathway for some of the pseudo-components present in the sample. Using the average effective activation energies for all components which are presented in Table 3, the SEAER value (Eq. (12)) was calculated. The value that was obtained is SEAER = 0.319. The actual result suggests that the effective activation energy of the UF-resin adhesive is lower than the expected E_a in the system with other three components.

Table 3. The CPRM parameters for pyrolysis process of the MDF derived Beech sawdust

β	A_l (min^{-1}) [a]				$E_{a,l}$ ($kJ\ mol^{-1}$) [a]				n_l [a]				φ_l [a]			
(°Cmin^{-1})	Co1	Co2	Co3	Co4	Co1	Co2	Co3	Co4	Co1	Co2	Co3	Co4	Co1	Co2	Co3	Co4
5	9.63×10^{16}	1.58×10^{7}	2.22×10^{17}	2.77×10^{19}	181.30	97.00	205.44	222.01					0.25	0.18	0.45	0.12
10	1.97×10^{15}	2.41×10^{7}	2.86×10^{18}	3.20×10^{20}	177.98	97.40	216.06	229.55	8.60	1.12	2.45	1.30	0.28	0.20	0.48	0.04
20	7.74×10^{16}	1.05×10^{8}	5.44×10^{17}	3.80×10^{19}	180.22	100.50	209.66	226.40					0.29	0.23	0.47	0.01
Average	5.86×10^{16}	4.83×10^{7}	1.21×10^{18}	1.29×10^{20}	179.83	98.30	210.39	225.99	-	-	-	-	0.27	0.20	0.47	0.06

[a] l = 1, 2, 3, and 4 (Hemicelluloses (Co1); UF-resin adhesive (Co2); Cellulose (Co3); Lignin (Co4)).

So, the presence of adhesive in the tested lignocellulosic material has no additive behavior for most of the properties shown in Table 1. As a consequence, it follows that the adhesive compound does not produce the synergistic effect.

Then the question arises, to what extent, and to which pseudo-component UF adhesive has an impact. In accordance with the obtained kinetic parameters presented in Table 3, the *QFit* analysis for the pyrolysis rate curves was performed and results are shown in Figure 8 a), b) and c). The statistical analysis results related to the considered heating rate were designated by COD-I (5°C min^{-1}), COD-II (10°C min^{-1}), and COD-III (20°C min^{-1}).

It can be observed that very high *QFit* values together with all COD magnitudes were obtained for proposed CPRM (Figure 8 a), b) and c)). From presented results, we can see that the Co1 deconvolution curve related to hemicelluloses does not vary significantly with the heating rate, and maximum decomposition rate is in the range of the temperatures expected for its decomposition (290.48 – 314.62°C; these results are in excellent agreement with results presented in Figure 7, attached to occurrence of "shoulder" on the DTG curve). However, it can be noted that Co1 decomposition kinetics is characterized by unrealistic high reaction order (n_1 = 8.60, Table 3), but the estimated value is very optimal for simulation. On the other hand, the deconvoluted rate curves attached to cellulose decomposition (Co3) significantly vary with the heating rate, not so much laterally along the temperature axis, as this is expressed in the change in the maximum pyrolysis rates (this phenomenon follows an identical phenomenon encompassed by DTG curves in terms of variation of the peak height which decreased with heating rate (Figure 2); this can be consequence of combined effects of the heat transfer and mass transfer). The cellulose pyrolysis kinetics was characterized by the much lower reaction order (n_3 = 2.45, Table 3) than one identified in the case of hemicelluloses pyrolysis. In the case of lignin pyrolysis (Co4), its decomposition was characterized with a much flatter pyrolysis curve at all considered heating rates, because the process runs slower and in a very wide temperature range.

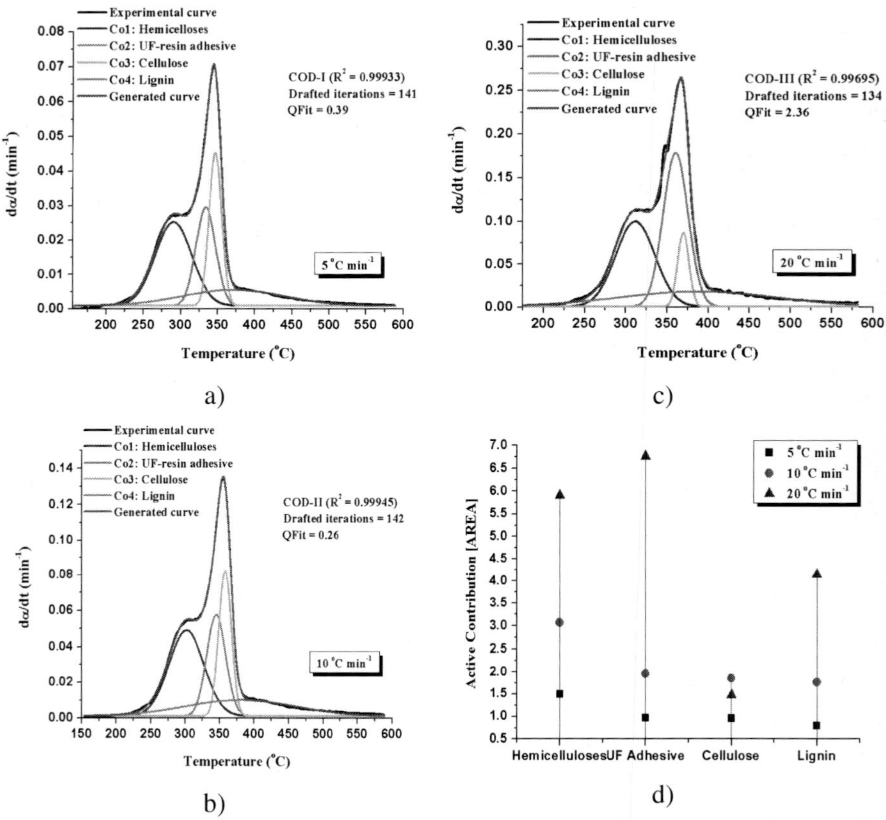

Figure 8. Deconvoluted pyrolysis rate curves of MDF derived Beech sawdust pyrolysis using 4-component model at different heating rates: a) 5°C min^{-1}, b) 10°C min^{-1}, and c) 20°C min^{-1}; d) Active contributions of all components during MDF derived Beech sawdust pyroysis in a function of applied heating rates.

It was found that the reaction order during lignin pyrolysis is equal to n_4 = 1.30 (Table 3) and this result is in agreement with results reported in the literature for lignin decomposition, where the reaction order can be different from the unity (Jiang et al. 2010). It was reported that the reaction order for the Klason lignin for both the softwood and hardwood species amounts about 1.50 (Jiang et al. 2010). For the thermal decomposition of lignin which is thermally induced, the E_a values can vary in a wide range as 120 – 361 kJ mol^{-1}. When hardwood lignin decomposed in comparison with softwood lignin, it was identified the increasing trend in pre-exponential factor (A) (which indicates the number of collisions between free radicals formed during lignin decomposition) (Poletto 2017) that leads to an increase of the effective activation energy values (Table 3).

However, an interesting phenomenon occurs in the case of UF-resin adhesive decomposition during pyrolysis. With an increase of heating rate, the pyrolysis rate of the UF-resin adhesive (Co2) rapidly increases and the complete rate curve moves near to the decomposition curve related to the cellulose. At the highest heating rate, the UF-resin adhesive pyrolysis rate curve "overrated" the cellulose pyrolysis rate curve (Figure 8 c)). At the same time, the cellulose decomposition curve decreases rapidly in the intensity. Based on these results, it can be seen that UF-resin adhesive directly affects the process of cellulose decomposition, during the pyrolysis of MDF derived Beech sawdust. Pyrolysis kinetics of UF-resin adhesive is very close to the first-order kinetics (n_2 = 1.12) and is significantly different from all other kinetic behaviors. It can be observed that the significant influence of the adhesive on the cellulose decomposition begins at a temperature around/and above 300 °C (Figure 8 a), b) and c)). This is reflected through the catalytic effect of the UF adhesive on the cellulose decomposition, and especially affected the release of small molecular compounds (such as CO and CO_2 gas molecules; Figure 4 d), e) and f)). Presence of UF adhesive may promote water formation and carboxylic acid during cellulose decomposition (Li et al. 2014). It can be assumed that UF adhesive participates in cellulose decomposition encouraging the chemical bonds breaking in cellulose molecule, where it can additionally stimulate and lignin decomposition. It should be noted that values of reaction orders different from the unity (Table 3) can be attributed to the presence of highly heterogeneous material, and also to the influence of the diffusion effects (which are not considered in Figure 7). It can be observed that the contribution of partial reactions to entire pyrolysis is different at various heating rates. This is primarily a consequence of material heterogeneous structure (see above), and because of such facts, the process with heterogeneous decomposition will be different at various heating rates. Figure 8 indicates that the pyrolysis process is very well described by proposed four parallel reaction models. However, the results presented for the highest heating rate (20°C min^{-1}) give some slightly poorer implementation if we consider *QFit* value. However, despite this, a satisfactory criterion for an estimated result at a given heating rate is still fulfilled.

The active contributions of all components during MDF derived Beech sawdust pyrolysis in a function of applied heating rates are shown in Figure 8 d). It can be observed that at the highest heating rate (20°C min^{-1}), the largest active contribution has the UF adhesive, while the smallest, at the same heating rate, has the cellulose. At the medium heating rate (10°C min^{-1}), the largest active contribution has the hemicelluloses. At the lowest heating rate (5°C min^{-1}), the active contributions from cellulose and UF adhesive are approximately the same (Figure 8 d)). However, an individual component, the cellulose shows the largest variation with the change in the heating rate. It can be assumed that adhesive species may affect the potential "partial inhibition" in respect to cellulose pyrolysis rate stifling. This effect is expressed at the higher temperatures. Therefore, the UF adhesive affects the reduction of the rate of cellulose pyrolysis, in particular, the achievement of its maximum value, which is in accordance with the previous discussion. However, the possible catalytic effect of the UF adhesive is most likely achieved in the early stages of cellulose decomposition where the various radicals were released. However, it is necessary to check the mentioned effects on the yield of released gaseous products at higher temperatures, which is determined based on the mass spectrometry analysis (Figure 4). Figure 9 a), b) and c) shows the gas yield productions (%) (CH_4, H_2 and CO_2) obtained at specific target temperatures, simultaneously monitored at different heating rates.

It can be seen from Figure 9 a) that the CH_4 production above 300°C is quite high beyond 60% (at all considered heating rates). At the higher temperatures, above 300°C, the production of H_2 is maximal and reaches 100%, at every heating rate (Figure 9 b)). For the production of CO_2, we have an interesting situation. At temperatures below 300°C, on the low and medium heating rates (5 and 10°C min^{-1}), the high production of CO_2 (~ 62.54%) is achieved. On the other hand, in the higher temperature region, above 300°C, the high production of CO_2 is achieved (~ 71.92%) only at the highest heating rate (Figure 9 c)).

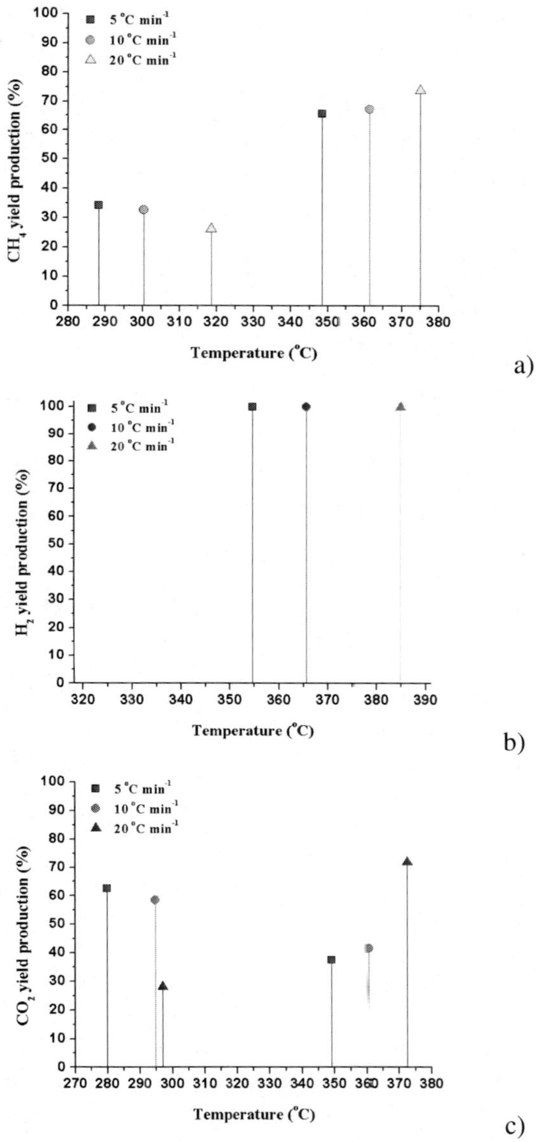

Figure 9. The gas yield productions (%) (CH_4 a), H_2 b) and CO_2 c)) obtained at target temperatures, simultaneously monitored at different heating rates.

In accordance with the results obtained, it can be pointed out that CH_4 and CO_2 production favored at low and medium heating rates and at temperatures below 300°C originates from breaking of the *methoxyl* group and the single C—C bonds. It can be assumed that CH_4 and CO_2 release by cracking of methoxyl groups, decarboxylation, and decarboxylation. Within the major pyrolysis stage, the releasing of CO_2 is a consequence of carboxyl breaking in the lateral chains of furfural acid as one of cellulose decomposition products (Zhao et al. 2017).

At much higher temperatures, beyond 400°C (not shown in Figure 9), the liberated volatiles undergo the secondary reactions which result in the formation of incondensable gases and of the small chain organic species. All these transformations originate from the lignin decomposition. Namely, there is a possibility of some influence of UF adhesive in that sense to instigate the ring-opening reaction in the lignin. On the other hand, the high H_2 production was obtained at the high temperatures, so the temperature rise will favor the hydrogen yields.

It should be noted that the water-gas shift reaction $CO + H_2O \rightarrow CO_2 + H_2$ can additionally increase the hydrogen production, where this reaction is quite pronounced in our studied case. In addition, the synergy in the gas yields during pyrolysis may be strongly related to the H_2 evolution. The H and OH radicals released in the pyrolysis process can promote maximization in the H_2 production in the temperature range which is indicated in Figure 9 b).

Reactivity Distribution Analysis

In this sub-section, we focus on gaining insight into the pyrolysis process of investigated lignocellulosic material by application of the distributed reactivity model (DRM) through multiple parallel reaction process, using analysis through the transition peaks for UF adhesive presence in the sample. Figure 10 shows experimentally obtained distribution curve of the effective activation energies for pyrolysis process of MDF derived Beech sawdust. The distribution curve shows the three distinct peaks related to the decomposition processes of the main pseudo-components: hemicelluloses (with maximum value of $E_{a,1}$ = 187.55 kJ mol^{-1}), cellulose (with maximum value of $E_{a,2}$ = 210.74 kJ mol^{-1}), and lignin (with maximum value of $E_{a,3}$ = 219.37 kJ mol^{-1}), respectively. However, from the experimental distribution curve, a separate peak which would be attributed to the decomposition of the adhesive compound was not identified.

The $f(E_a)$ curve is widespread with the distribution of E_a values in the range of 155.80 — 322.30 kJ mol^{-1}, where for the solid state reactions, the E_a values were between 50.00 and 350.00 kJ mol^{-1} (Cai et al. 2013). The latest fact confirms that the obtained experimental distribution curve shown in Figure 10 is quite reasonable.

Figure 11 shows 4-components distributed reactivity model with the Log-Normal distribution of effective activation energies $f_p(E_a)$ (p = 1, 2, 3 and 4) related to the decomposition reactions of all presented reactive components in studied material. The model simulation was successfully implemented with a simulation quality of 100% (Simulation Status = 100%). Table 4 lists the values of distribution parameters obtained for each individual distribution, relating to the decomposition reactions of the observed components.

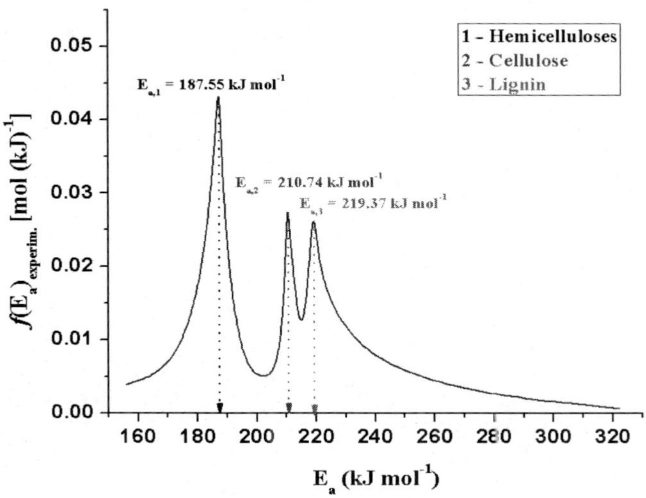

Figure 10. The experimental distribution curve of the effective activation energies for pyrolysis process of MDF derived Beech sawdust.

Table 4. The 4-components Log-normal distribution reactivity model for MDF derived Beech sawdust pyrolysis

p = 1, Component 1 – UF-resin adhesive		
$f_{0,1}$ **(kJ mol^{-1})**	σ_1 (kJ mol^{-1})	μ_1 (kJ mol^{-1}) [a]
0.16	0.06	168.83
p = 2, Component 2 - Hemicelluloses		
$f_{0,2}$ **(kJ mol^{-1})**	σ_2 (kJ mol^{-1})	μ_2 (kJ mol^{-1}) [a]
0.72	0.03	189.37
p = 3, Component 3 - Cellulose		
$f_{0,3}$ **(kJ mol^{-1})**	σ_3 (kJ mol^{-1})	μ_3 (kJ mol^{-1}) [a]
0.36	0.01	220.07
p = 4, Component 4 - Lignin		
$f_{0,4}$ **(kJ mol^{-1})**	σ_4 (kJ mol^{-1})	μ_4 (kJ mol^{-1}) [a]
0.55	0.04	221.89

[a] $E_{a,p}$ values originate from estimated distribution curves in Figure 11.

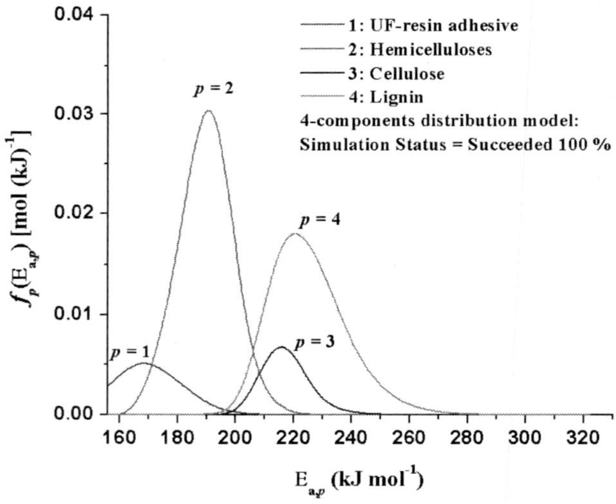

Figure 11. The four (-4)—components distributed reactivity model with the Log-Normal distribution of effective activation energies fp(Ea) (p = 1, 2, 3 and 4) related to decomposition reactions of all reactive components presented in studied lignocellulosic material.

It can be observed from Table 4 that $f_{0,1} < f_{0,3} < f_{0,4} < f_{0,2}$ which suggests that smallest participation in bond breaking degrees belongs to UF-resin adhesive, while largest participation can be attributed to hemicelluloses.

However, the distribution curve related to adhesive decomposition is shifted strictly to lower E_a values, but with much broader $f_1(E_{a,1})$ profile ($\sigma_1 > \sigma_4 > \sigma_2 > \sigma_3$ (Table 4)) (Figure 11). It can be noted that based on estimated model, the cellulose decomposition occurs in the narrow E_a range, as well as a narrower temperature range (Table 4). However, the mean value μ_1 = 168.83 kJ mol^{-1} is within the range of UF resen decomposition (150 – 300 kJ mol^{-1}) and this value is higher than the one identified in the case of CPRM (98.30 kJ mol^{-1} (Table 3)). The value of SEAER obtained for this case amounted SEAER = 0.535. So, SEAER (4-components distributed reactivity model) > SEAER (CPRM), but it is still far below the unity. Consequently, the synergistic effect does not exist, where this result confirms previous findings. The UF adhesive is characterized by high reactivity (lower E_a's than those for other components) with apparent symmetric behavior (Figure 11 and Table 4), where in the case of lower content of formaldehyde may consist a narrow range of reactions during decomposition. This can be expected since that UF polymer has a relatively simple structure, which means a less variety in the type of the bonds present. Namely, the UF resin adhesive may gradually decompose out volatile compound such as free formaldehyde in a temperature range of 200 – 330°C. This probably due to the pyrolysis of the cured UF accompanies with formation of diureidomethane, hydroxymethyl and formaldehyde evolution (Hong et al. 2016). The estimated $f_1(E_{a,1})$ distribution curve with associated E_a spreading are in accordance with literature reports attached to UF resin adhesive decomposition (125.82 — 246.96 kJ mol^{-1}) (Hong et al. 2016). The resulting facts are interesting because actual results are more realistic in the context of synergistic effects. Namely, in this case, the expected temperature range for UF resin adhesive decomposition is approximately ahead of the temperature threshold for the hemicelluloses decomposition and ends approximately where it would end the same process. However, this phenomenon cannot be seen on DTG curves as the appearance of another peak or the emergence of an additional "shoulder" that would originate from UF resin adhesive in above-indicated temperature region (see Figure 2). This is distinguished from results that arise from CPRM, but however, the latest results are in full co-existence since that two approaches strongly confirm the deviation from additivity. Unlike the hemicelluloses distribution curve $f_2(E_{a,1})$, other distributions show certain Log-Normal asymmetric behavior (Figure 11). This asymmetry arises from a sum of mutually correlated lognormal and as a consequence, the asymmetry in the upper tails occurs (Figure 11). Also, this asymmetry can be explained from the much wider variety of bonds present for the formation of the radical reactions. However, $f_{0,2}$ value for hemicelluloses is the highest in relation to other f_0 values (Table 4), and this means that under the given conditions, hemicelluloses exhibit a high capacity in the tendency of an easier bond breaking, but this is not directly linked to the variety of bonds. In comparison with the value of $f_{0,3}$ for cellulose (Table 4) and more asymmetrical distribution (Figure 11), the pyrolysis mechanism of these components are probably different, and that would be reflected through the consideration of different E_a and n values, whether unzipping starts at the end of chain or elsewhere, and if the levoglucosan is formed as an intermediate. Generally, the three emulous cellulose decomposition pathways occur, such as 1) the depolymerization to the levoglucosan and other primary anhydrosugars, hereafter, 2) the fragmentation to the hydroxyacetaldehyde and carbonyls, acids, and alcohols, and 3) dehydration to the bio-char, gases (CO_2, CO), and water. Dehydration reaction is governed predominantly at low temperatures and slower heating rates. At the higher temperatures, the depolymerization and fragmentation take place. Taking into consideration the distribution $f_{0,3}$ for cellulose decomposition, the reported E_a value linked to the formation of "active cellulose" (242.80 kJ mol^{-1}) (for cellulose pyrolysis) (Shen et al. 2011) sits on the right–tail of distribution at higher E_a values (the target temperature range for this step: $\Delta T = 300 - 450^{\circ}$C). At higher temperatures (optimal at about 600°C) there is a fragmentation, which is followed by a gradual decline in the effective activation energy. The primary volatiles liberation attached to gases production takes place with approximately $E_a \sim 198$ kJ mol^{-1} (Shen et al. 2011), located on the left–tail of the distribution obtained for cellulose decomposition, during the MDF derived Beech sawdust pyrolysis. In a rather narrow range of temperatures and E_a values, water is released, as well as the CO and CO_2 gases required in water-gas shift reaction. With further temperature elevating, together with H_2 releasing, the bio-char process occurs. Graphical presentation of the mechanism is shown in Figure 12.

Table 5. The parameters of 3-components (hemicelluloses, cellulose, and lignin) distributed reactivity model at various heating rates established for the pyrolysis processs of MDF derived Beech sawdust

Condition parameters [a]	Hemicelluloses (Pseudo-component 1)	Cellulose (Pseudo-component 2)	Lignin (Pseudo-component 3)
β (°C min^{-1})	5	5	5
w_p	0.19	0.65	0.16
μ_p (kJ mol^{-1})	189.37	220.07	221.89
σ_p (kJ mol^{-1})	0.03	0.01	0.04
n_p	8.45	2.86	1.38
A_p (min^{-1})	1.30×10^{17}	0.50×10^{19}	6.01×10^{20}
Obj.Func.	3.55	1.74	0.85
β (°C min^{-1})	10	10	10
w_p	0.22	0.59	0.19
μ_p (kJ mol^{-1})	189.37	220.07	221.89
σ_p (kJ mol^{-1})	0.03	0.01	0.04
n_p	8.45	2.86	1.38
A_p (min^{-1})	1.68×10^{18}	5.10×10^{19}	3.50×10^{20}
Obj.Func.	3.79	1.65	0.95
β (°C min^{-1})	20	20	20
w_p	0.16	0.63	0.21
μ_p (kJ mol^{-1})	189.37	220.07	221.89
σ_p (kJ mol^{-1})	0.03	0.01	0.04
n_p	8.45	2.86	1.38
A_p (min^{-1})	4.21×10^{18}	1.05×10^{19}	0.60×10^{21}
Obj.Func.	3.98	1.80	1.11

[a]Differential data.

$$ObjectiveFunc\left(A_p,\mu_p,\sigma_p\right)=100\cdot\left(\frac{d\alpha}{dt}\right)_{\max}\sqrt{\frac{\sum_{k=1}^{N_k}\left[\frac{d\alpha_p\left(T\middle|A_p,\mu_p,\sigma_p\right)}{dt}-\left(\frac{d\alpha_p(T)}{dt}\right)_{\exp}\right]^2}{N_k}},$$

where (dαp(T)/dt)exp is the actual rate curve for the considered pseudo-component decomposition, (dα/dt)max is the maximum rate value, while dαp(T|Ap,μp,σp)/dt is the computed (weighted) decomposition rate curve related to the p-th pseudo-component (for p = 1, 2, 3); N is the number of data points.

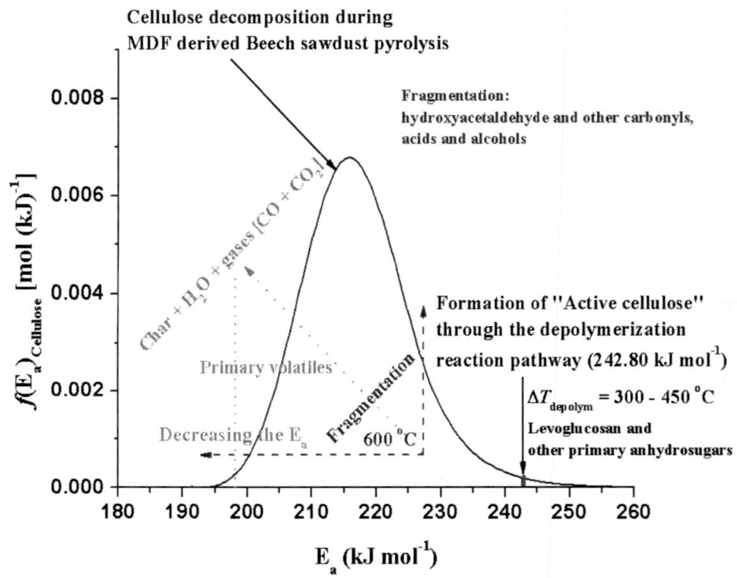

Figure 12. Graphical presentation of cellulose decomposition in pyrolytic process of MDF derived Beech sawdust.

In accordance with the previous discussion (including specific temperature regions), decarboxylation and decarbonylation processes take place that produces evolving H_2O, CO, and CO_2 (Figure 4 d), e) and f)). In addition, the lignin decomposition starts later (considering μ-values (Figure 11 and Table 4)) than hemicelluloses and cellulose contributions and continues until the end of the pyrolysis process. This can be expected since that lignin consists phenyl ether bonds that must be broken to give phenols and benzyl compounds with carboxylic acid functional groups.

The comparison between actual pyrolysis rate curves and computed (weighted) curves applying distributed reactivity model was performed. For estimation of weighted rate curves, the corresponding kinetic parameters were calculated and presented in Table 5.

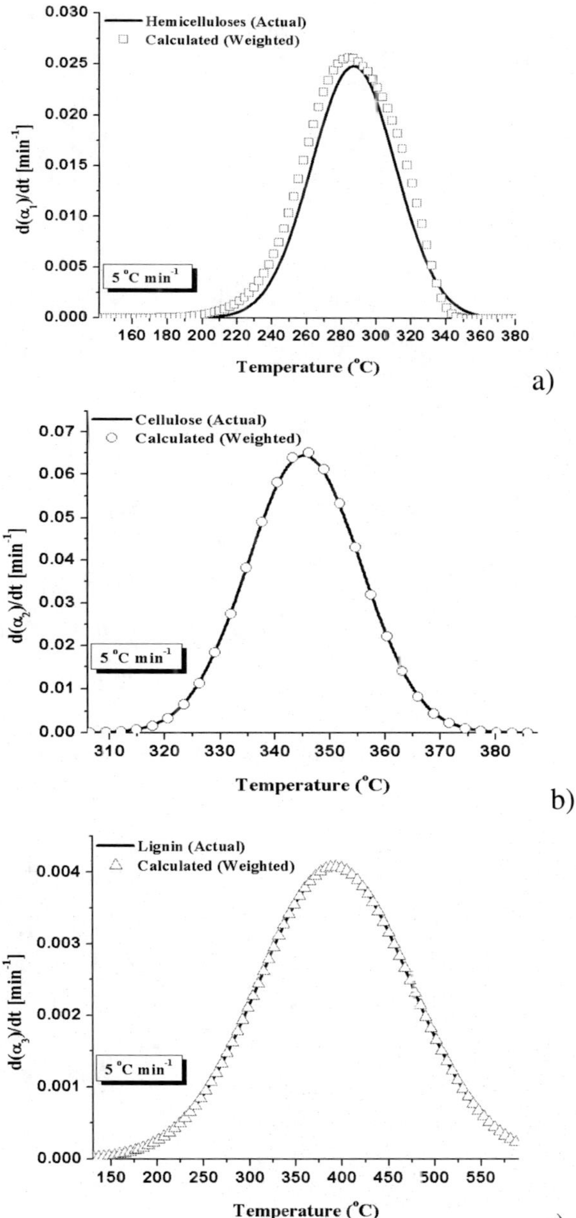

Figure 13. Comparison between actual and weighted pyrolysis rate curves for three (-3) component Log-Normal distributed reactivity model at $\beta = 5^{o}C\ min^{-1}$, excluding the adhesive component.

Figure 13 a), b) and c) shows the comparison between actual and weighted pyrolysis rate curves for 3-component Log-Normal distributed reactivity model, estimated at $\beta = 5^{o}C\ min^{-1}$.

The weighted curves show the qualitative contributions of each of pseudo-components (1: hemicelluloses, 2: cellulose, and 3: lignin) to the entire pyrolysis process, as well as their active effective activation energy distributions. The peak of the rate curves, from hemicelluloses up to lignin (observed at the other heating rates) is shifted to the higher temperatures, prolonging the decomposition temperature ranges. The model weighted curves very well describe the actual rate pyrolysis curves, for estimated sets of kinetic parameters which are shown in Table 5. The corresponding weighted curve for hemicelluloses pyrolysis shows certain departure (also at the other heating rates) reflecting the somewhat lower quality of fit (Table 5) (however, it is considered that real-based kinetic mechanism described by a given model can have a quality of fit (Objective function) value up to 10%). The four (-4) components contribute to the major dimensionality of the MDF derived Beech sawdust pyrolysis, and undeniably represent the entire kinetic mechanism of the process. However, regardless of this, on the basis of above-presented results, the pyrolysis process could also be described with three (-3) components (namely, the pseudo-components of biomass sample) with respect to the existence of a strong deviation from additive behavior (no overlaps of E_a distribution peaks between UF adhesive and biomass pseudo-components), so indicates the absence of chemically interactive synergistic effects. Certain small differences in the values of kinetic parameters observed in CPRM and the Log-Normal distributed reactivity approaches is a consequence of the application of the multi-component kinetic analysis, which uses a variety of components with higher pre-exponential factors.

CONCLUSION

A novel kinetic model for non-isothermal pyrolysis of *MDF (medium-density fiberboard)* derived Beech sawdust was proposed. The pyrolysis process was monitored using the simultaneous thermal analysis (STA) coupled with mass spectrometry (MS) analysis. The chemical analysis showed that raw material has desirable characteristics in thermal decomposition regime as an initial step in combustion processing, as well as exhibits favorable properties for the production of activated carbons. Comparing with other systems, it was shown that the MDF derived Beech sawdust belongs to lignocellulosic material with increased cellulose content. It was found that various rate of the heating can favor different reaction stages in the pyrolysis process of collapsed material, as low heating rates favor the dehydration to the bio-char, while the high heating rates favor the depolymerization to the levoglucosan. The influence of the process parameters such as heating rate, temperature and time, on the TG (thermogravimetry) pyrolysis profiles was tested through matrix analysis. Based on the current analysis, it was found that for quality pyrolysis performing, the lower values of the heating rates (where the possibility of the secondary reactions is significantly reduced) are highly desirable. From the behavior of thermo-analytical curves, it was established that the presence of UF adhesive in MDF derived Beech sawdust affects "slows down" the decomposition of cellulose in an active pyrolysis zone, and this is probably reflected in a suppression of the rate of pyrolysis. A detailed analysis of the release of the gaseous products that arise from the decomposition of the pseudo-components was carried out. It was found that CH_4 and CO_2 production is favored at low and medium heating rates and at temperatures below 300°C, originates from breaking of the *methoxyl* group and single C—C bonds. At much higher temperatures (above 400°C), it was established that liberated volatiles undergoes secondary reactions which result in the formation of incondensable gases and of the small chain organic species. This was discussed as a transformation that originated from lignin decomposition.

The new proposed model which combines two approaches as combined four parallel reaction model (CPRM) and three-component Log-Normal distributed reactivity model, was successfully described the kinetics of the pyrolysis process. The CPRM approach showed that with an increase of heating rate, the pyrolysis rate of UF-resin adhesive rapidly increases and its rate curve moves very close to decomposition curve related to the cellulose. At the highest heating rate, the UF-resin adhesive pyrolysis rate curve "overrated" cellulose–related curve. It was assumed that UF adhesive participates in cellulose decomposition, encouraging the chemical bonds breaking in cellulose molecule, where it can additionally stimulate and lignin decomposition. However, based on the analysis of the synergy effective activation energy ratio (SEAER), it was found that the presence of an adhesive compound in studied material has no additive behavior for most of its (chemical) properties.

The three-component Log-Normal distributed reactivity model was confirmed that synergistic effect does not exist. It was found that certain asymmetries in the distribution curves of the effective activation energies for some pseudo-components are a consequence of the wider variety of bonds present and/or "capacity" in the tendency of an easier bond breaking. It was found that certain small differences in the values of kinetic parameters observed in both approaches, is a consequence of the application of the multi-component kinetic analysis, which uses a variety of components.

ACKNOWLEDGMENTS

Authors would like to acknowledge financial support of Ministry of Education, Science and Technological Development of the Republic of Serbia under the Projects 172015 and III42010.

REFERENCES

Ahamad, T., Alshehri, S. M. (2014). Thermal degradation and evolved gas analysis: A polymeric blend of urea formaldehyde (UF) and epoxy (DGEBA) resin. *Arabian Journal of Chemistry* 7:1140-1147.

Aigbomian, E. P., Fan, M. (2013). Development of wood-crete from hardwood and softwood sawdust. *The Open Construction & Building Technology Journal* 7:108-117.

Aqsha, A., Mahinpey, N., Mani, T., Salak, F., Murugan, P. (2011). Study of sawdust pyrolysis and its devolatilisation kinetics. *The Canadian Journal of Chemical Engineering* 89(6):1451-1457.

Antal, Jr M. J. (1983). Effects of reactor severity on the gas-phase pyrolysis of cellulose- and kraft lignin-derived volatile matter. *Industrial & Engineering Chemistry Product Research and Development* 22(2):366-375.

Antal, Jr M. J., Várhegyi, G. (1995). Cellulose pyrolysis kinetics: The current state of knowledge. *Industrial & Engineering Chemistry Research* 34(3):703-717.

Aviles, J. P., Brown, R. C. (2014). The Effect of Heating rate on the Conversion of Cellulose to Levoglucosan and Char. Conference Paper, TCS 2014: *Symposium on Thermal and Catalytic Sciences for Biofuels and Biobased Products,* Denver, Colorado, DOI: 10.13140/2.1.2294.4961.

Avrami, M. J. (1939). Kinetics of phase change. I. General theory. *Chemical Physics* 7:1103-1112.

Basu, P. (2010). *Biomass Gasification and Pyrolysis: Practical Design and Theory.* 1st Edition, ISBN: 9780080961620, Academic Press, Elsevier, Burlington, MA 01803, USA, pp. 1-364.

Bodîrlậu, R., Spiridon, I., Teacậ, C. A. (2007). Chemical investigation of wood tree species in temperate forest in East-Northern Romania. *BioResources* 2(1):41-57.

Bradbury, A. G. W., Sakai, Y., Shafizadeh, F. (1979). A kinetic model for pyrolysis of cellulose. *Journal of Applied Polymer Science* 23: 3271-3280.

Brebu, M., Vasile, C. (2010). Thermal degradation of lignin – A review. *Cellulose Chemistry and Technology* 44:353-363.

Bridgwater, A. V., Toft, A., Brammer, J. (2002). A techno-economic comparison of power production by biomass fast pyrolysis with gasification and combustion. *Renewable and Sustainable Energy Reviews* 6(3):181-246.

Bridgwater, A. V. (2003). Renewable fuels and chemicals by thermal processing of biomass. *Chemical Engineering Journal* 91:87-102.

Bridgwater, A. V. (2009). Biomass Pyrolysis, In: *Thermal Biomass Conversion.* Bridgwater AV, Hofbauer H, and Van Loo S (Eds.), ISBN: 9781872691534 1872691536, CPL Press, Newbury, UK, pp. 37-78.

Burkhardt, S. J. (2013). *Forest Residues as a Potential Feedstock for a Biorefinery: Material Balance and Pretreatment Strategies.* Thesis submitted for the degree of Master of Applied Science in The Faculty of Graduate and Postdoctoral Studies (Forestry), The University of British Columbia (Vancouver), Canada, August 2013, pp. 1-99.

Cai, J., Wu, W., Liu, R., Huber, G. W. (2013). A distributed activation energy model for the pyrolysis of lignocellulosic biomass. *Green Chemistry* 15:1331-1340.

Cao, Y., Wang, Y., Riley, J. T., Pan, W. P. (2006). A novel biomass air gasification process for producing tar-free higher heating value fuel gas. *Fuel Processing Technology* 87:343-353.

Channiwala, S. A., Parikh, P. P. (2002). A unified correlation for estimating HHV of solid, liquid and gaseous fuels. *Fuel* 81(8):1051-1063.

Chen, W. H., Hsu, H. C., Lu, K. M., Lee, W. J., Lin, T. C. (2011). Thermal pretreatment of wood (Lauan) block by torrefaction and its influence on the properties of the biomass. *Energy* 36:3012-3021.

Chen, S., Li, S., Mu, J., Feng, Y. (2015). Influence of urea formaldehyde resin on the pyrolysis characteristics and gas evolution of waste MDF. *Wood Research* 60(1):113-124.

Cotana, F., Cavalaglio, G., Gelosia, M., Nicolini, A., Coccia, V., Petrozzi, A. (2014). Production of bioethanol in a second generation prototype from pine wood chips. *Energy Procedia* 45:42-51.

Demirbaş, A. (2001). Biomass resource facilities and biomass conversion processing for fuels and chemical. *Energy Conversion and Management* 42:1357-1378.

Di Blasi, C. (2008). Modeling chemical and physical processes of wood and biomass pyrolysis. *Progress in Energy and Combustion Science* 34:47-90.

Doyle, C. D. (1962). Estimating isothermal life from thermogravimetric data. *Journal of Applied Polymer Science* 6:639-642.

EN 622-5 (2006). Fibreboards – Specifications. Part 5. *Requirements for dry process boards (MDF),* European Committee for Standardization (CEN), Brussels, Belgium.

EN ISO 17225-1 (2014). *Solid biofuels – Fuel specifications and classes. Part 1.* General requirements, European Committee for Standardization (CEN), Brussels, Belgium.

EN ISO 16948 (2015). *Solid biofuels – Determination of total content of carbon, hydrogen and nitrogen,* European Committee for Standardization (CEN), Brussels, Belgium.

EN ISO 18134 (2015). *Solid biofuels – Determination of moisture content – Oven dry method.* Part 1. Total moisture – Reference method, European Committee for Standardization (CEN), Brussels, Belgium.

EN ISO 14780 (2017). *Solid biofuels – Sample preparation,* European Committee for Standardization (CEN), Brussels, Belgium.

EN ISO 18125 (2017). *Solid biofuels – Determination of calorific value, European Committee for Standardization (CEN),* Brussels, Belgium.

Farias, T. M., Cardozo, N. S. M., Secchi, A. R. (2013). The use of Gauss-Hermite quadrature in the determination of the molecular weight distribution of linear polymers by rheometry. *Brazilian Journal of Chemical Engineering* 30(4):909-921.

Ferdous, D., Dalai, A. K., Bej, S. K., Thring, R. W., Bakhshi, N. N. (2001). Production of H_2 and medium Btu gas via pyrolysis of lignins in a fixed-bed reactor. *Fuel Processing Technology* 70:9-26.

Foster, C. E., Martin, T. M., Pauly, M. (2010). Comprehensive compositional analysis of plant cell walls (Lignocellulosic biomass) Part II: Carbohydrates. *Journal of Visualized Experiments* 37:e1837 doi:10.3791/1837.

Grønli, M. G., Várhegyi, G., Di Blasi, C. (2002). Thermogravimetric analysis and devolatilization kinetics of wood. *Industrial & Engineering Chemistry Research* 41:4201-4208.

Gu, J., Catchmark, J. M. (2013). The impact of cellulose structure on binding interactions with hemicellulose and pectin. *Cellulose* 20(4):1613-1627.

Hong, W., Meng, M., Gao, D., Liu, Q., Kang, C., Huang, S., Zhou, Z., Chen, C. (2016). Thermal analysis study of modified urea-formaldehyde resin. *Polymer* (*Korea*) 40(5):707-713.

Hosoya, T., Kawamoto, H., Saka, S. (2007). Cellulose–hemicellulose and cellulose–lignin interactions in wood pyrolysis at gasification temperature. *Fuel* 80:118-125.

Isikgor, F. H., Becer, C. R. (2015). Lignocellulosic biomass: A sustainable platform for the production of bio-based chemicals and polymers. *Polymer Chemisty* 6:4497-4559.

Jakab, E., Faix, O., Till, F. (1997). Thermal decomposition of milled wood lignins studied by thermogravimetry/mass spectrometry. *Journal of Analytical and Applied Pyrolysis* 40-41:171–186.

Jiang, G., Nowakowski, D. J., Bridgwater, A. V. (2010). A systematic study of the kinetics of lignin pyrolysis. *Thermochimica Acta* 498:61-66.

Joshua, J. A., Ahiekpor, J. C., Kuye, A. (2016). Nigerian hardwood (*Nesogordonia papaverifera)* sawdust characterization: Proximate analysis, cellulose and lignin contents. *Lignocellulose* 5(1):50-58.

Karaosmanoğlu, F., Tetik, E., Göllü, E. (1999). Biofuel production using slow pyrolysis of the straw and stalk of the rapeseed plant. *Fuel Processing Technology* 59:1-12.

Kilzer, F. J., Broido, A. (1965). Speculation on the nature of cellulose pyrolysis. *Pyrodynamics* 2:151-163.

Klein, M. T., Virk, P. S. (2008). Modeling of lignin thermolysis. *Energy & Fuels* 22(4):2175-2182.

Lacovidou, E., Hakladakis, J., Deans, I., Velis, C., Purnell, P. (2018). Technical properties of biomass and solid recovered fuel (SRF) co-fired with coal: Impact on multi-dimensional resource recovery value. *Waste Management* 73:535-545.

Lai, Z., Li, S., Zhang, Y., Li, Y., Mu, J. (2018). Influence of urea formaldehyde resin on the pyrolysis of biomass components: Cellulose, hemicellulose, and lignin. *BioResources* 13(2):2218-2232.

Lamaming, J., Sulaiman, O., Sugimoto, T., Hashim, R., Said, N., Sato, M. (2013). Influence of chemical components of oil palm on properties of binderless particleboard. *BioResources* 8(3): 3358-3371.

Li, S., Chen, S., Mu, J. (2012). Influence of UF resin on pyrolysis characteristics of biomass components: A thermogravimetric study. *Proceedings of 2012 International Conference on Biobase Material Science and Engineering (BMSE)*, October 21-23, 2012, Changsha, China, pp. 1-5 DOI: 10.1109/BMSE.2012.6466171.

Li, S., Mu, J., Zhang, Y. (2014). Influence of urea formaldehyde resin on pyrolysis of biomass: A modeling study by TG-FTIR. *Spectroscopy and Spectral Analysis* 34(6):1497-1501.

Mamleev, V., Bourbigot, S., Le Bras M., Yvon, J., Lefebvre, J. (2006). Model-free method for evaluation of activation energies in modulated thermogravimetry and analysis of cellulose decomposition. *Chemical Engineering Science* 61:1276-1292.

Mamleev, V., Bourbigot, S., Yvon, J. (2007). Kinetic analysis of the thermal decomposition of cellulose: The change of the rate limitation. *Journal of Analytical and Applied Pyrolysis* 80:141-150.

Mishra, R. K., Mohanty, K. (2018). Pyrolysis kinetics and thermal behavior of waste sawdust biomass using thermogravimetric analysis. *Bioresource Technology* 251:63-74.

Mohan, D., Pittman, Jr C. U., Steele, P. H. (2006). Pyrolysis of wood/biomass for bio-oil: A critical review. *Energy & Fuels* 20:848-889.

Moomaw, W., Yamba, F. D., Kamimoto, M., Wiser, R. (2012). Renewable Energy and Climate Change, In: *PCC Special Report on Renewable Energy Sources and Climate Change Mitigation,* Ch. 1, Edenhofer O, Pichs R, Sokona Y, Seyboth K, Matschoss P, Kadner S, Zwickel T,

Eickemeier P, Hansen G, Schlömer S, Stechow C (Eds.), Cambridge University Press, Cambridge, UK, pp. 161-207.

Mu, J., Yu, Z., Zhang, D., Jin, X. (2011). Pyrolysis characteristics of disused composite panels *and* properties of its products. *Journal of Beijing Forestry University 33(1):125-128.*

Nanda, S., Dalai A. K., Kozinski, J. A. (2013). Forestry biomass in a bioenergy perspective. *Journal* of *Science & Technology for Forest Products and Processes* 3(6):15-26.

Ozawa, T. A. (1965). A new method of analyzing thermodynamic data. *Bulletin of the Chemical Society of Japan* 38:1881-1886.

Pasangulapati, V., Kumar, A., Jones, C. L., Huhnke, R. L. (2012). Characterization of switchgrass, cellulose, hemicellulose and lignin for thermochemical conversions. *Journal of Biobased Materials and Bioenergy* 6: 249-258.

Pereira, B. L. C., Carneiro, A. C. O., Carvalho, A. M. M. L., Colodette, J. L., Oliveira, A. C., Fontes, M. P. F. (2013). Influence of chemical composition of Eucalyptus wood on gravimetric yield and charcoal properties. *BioResources* 8(3):4574-4592.

Pereira, R. G., Martins, C., Mendes, N., Farias, L., Ferreira, R. C., Oliveira, A., Oliveira, M., da Costa, R. (2014). Preparation of activated carbons from cocoa shells and siriguela seeds using H_3PO_4 and $ZnCl_2$ as activating agents for BSA and α-lactalbumin adsorption. *Fuel Processing Technology* 126:476-486.

Pérez-Maqueda, L. A., Criado, J. M. (2000). The accuracy of Senum and Yang's approximations to the Arrhenius integral. *Journal of Thermal Analysis and Calorimetry* 60(3):909-915.

Phanphanich, M., Mani, S. (2011). Impact of torrefaction on the grindability and fuel characteristics of forest biomass. *Bioresource Technology* 102:1246-1253.

Poletto, M. (2017). Assessment of the thermal behavior of lignins from softwood and hardwood species. *Maderas. Ciencia y Tecnologia* 19(1):63-74.

Saarela, K.-E., Harju, L., Rajander, J., Lill, J.-O., Heselius, S.-J., Lindroos, A., Mattson, K. (2005). Elemental analyses of pine bark and wood in an environmental study. *Science of the Total Environment* 343(1-3):231-234.

Savova, D., Apak, E., Ekinci, E., Yardim, F., Petrov, N., Budinova, T., Razvigorova, M., Minkova, V. (2001). Biomass conversion to carbon adsorbents and gas. *Biomass and Bioenergy* 21:133-142.

Schuerch, C. (1967). Methods of Wood Chemistry. Vol. II., Browning BL (Ed.), Wiley, New York, pp. 498, Publication 1968: *J Polym Sci Part A-2 Polym Phys* 6(11) pp 1943-1944.

Shen, D., Xiao, R., Gu, S., Luo, K. (2011). The pyrolytic behavior of cellulose in lignocellulosic biomass: A review. *RSC Advances* 1:1641-1660.

Senneca, O. (2007). Kinetics of pyrolysis, combustion and gasification of three biomass fuels. *Fuel Processing Technology* 88(1):87-97.

Şensöz, S. (2003). Slow pyrolysis of wood barks from *Pinus brutia* Ten. and product compositions. *Bioresource Technology* 89:307-311.

Stefanidis, S. D., Kalogiannis, K. G., Iliopoulou, E. F., Michailof C. M., Pilavachi, P. A., Lappas, A. A. (2014). A study of lignocellulosic biomass pyrolysis via the pyrolysis of cellulose, hemicellulose and lignin. *Journal of Analytical and Applied Pyrolysis* 105:143-150.

TAPPI T 222 om-02 (2006). *Acid-insoluble lignin in wood and pulp* (accessed January 8, 2018).

Tarrés, Q., Pellicer, N., Balea, A., Merayo, N., Negro, C., Blanco A., Delgado-Aguilar, M., Mutjé, P. (2017). Lignocellulosic micro/nanofibers from wood sawdust applied to recycled fibers for the production of paper bags. *International Journal of Biological Macromolecules* 105:664-670.

Vamvuka, D., Kakaras, E., Kastanaki, E., Grammelis, P. (2003). Pyrolysis characteristics and kinetics of biomass residuals mixtures with lignite. *Fuel* 82(15-17):1949-1960.

Van de Velden, M., Baeyens, J., Brems, A., Janssens, B., Dewil, R. (2010). Fundamentals, kinetics and endothermicity of the biomass pyrolysis reaction. *Renewable Energy* 35:232-242.

Vassilev, S. V., Baxter, D., Andersen, L. K., Vassileva, C. G. (2010). An overview of the chemical composition of biomass. *Fuel* 89(5):913-933.

Vyazovkin, S., Sbirrazzuoli, N. (1997). Confidence intervals *for the* activation energy estimated *by* few experiments. *Analytica Chimica Acta 355:175-180.*

Vyazovkin, S. (2001). Modification of the integral isoconversional method to account for variation in the activation energy. *Journal of Computational Chemistry* 22:178-183.

Wang, G., Li, W., Li, B., Chen, H. (2008). TG study on pyrolysis of biomass and its three components under syngas. *Fuel* 87:552-558.

Widyawati, M., Church, T. L., Florin, N. H., Harris, A. T. (2011). Hydrogen synthesis from biomass pyrolysis with in situ carbon dioxide capture using calcium oxide. *International Journal of Hydrogen Energy* 36:4800-4813.

Willner, T., Brunner, G. (2005). Pyrolysis kinetics of wood and wood components. *Chemical Engineering & Technology* 28:1212-1225.

Yang, G., Jaakkola, P. (2011). *Wood Chemistry and Isolation of Extractives from Wood.* BIOTULI Project, Saimaa University of Applied Sciences, Lappeenranta and Imatra, South Karelia, Finland, pp. 3-47.

Zhang, Y., He, Z. B., Xue, L., Chu, D. M., Mu, J. (2016). Influence of a urea-formaldehyde resin adhesive on pyrolysis characteristics and volatiles emission on poplar particleboard. *RSC Advances* 6:12850-12861.

Zhao, C., Jiang, E., Chen, A. (2017). Volatile production from pyrolysis of cellulose, hemicellulose and lignin. *Journal of the Energy Institute* 90(6):902-913.

INDEX

A

B

C

D

E

F

G

H

I

K

L

M

N

O

P

R

S

T

U

V

W

X

Y